古陽建境

龙 门 博 物 馆 建 筑 设 计

Architectural Realm of Guyang
THE LONGMEN MUSEUM ARCHITECTURE

邹迎晞 编著
Zou Yingxi

中国书店

项目名称: 龙门博物馆
Project name: Longmen Museum

时间: 2006年

建筑设计师:邹迎晞、丹尼尔[德]

Time: 2006

Architects: Zou Yingxi, Daniel Schwabe

建筑设计公司:

Architectural design company:

SYN建筑师事务所
SYN architect office

北 京 龙 安 华 诚 建 筑 设 计 有 限 公 司
Beijing Longanhuacheng Architectural Designing Co.,Ltd.

内容简介

《古阳建境》是记录龙门博物馆历经7年设计、研讨及建设过程的，关于设计理念与思路的建筑设计专题图书。书中特别就建筑设计理念的思考过程，从前期准备，到初步确定，以及理念分项内容作了较详细的阐述。并就设计理念的概念与意义作了分析。最后就龙门博物馆设计的思考与中国现代建筑的发展作出探索性思考。

作者希望能以本书的出版，在自身于中国传统哲学思想的从被动到主动的借鉴过程中，提示建筑设计行业的广大同行，现代中国建筑的发展应立足本民族的文化，重视地域性，并于建筑师的主观性与设计对象之间的关系作深入研究，尊重普遍性的基础上，发挥个性与艺术性。

努力创造中国自己的建筑设计道路。就此问题，本书还就中国传统艺术中的"情境"，与中国的"心性与直觉"及"有无相生"等思维观作了描述。

本书除可供建筑设计从业人员参考外，也适合高等院校、建筑院所等机构的人员参考。

目 录
CONTENTS

前言

天造伊阙，禹开龙门。文明光炽，精神恒长。九皋瑞鸟声远，陆浑清漪泱泱。两山一水,神秀钟于斯地。天人合一,成就大块文章。更佛法东传，根植于洛阳! 龙门石窟，十万造像，万丈佛光。曹衣出水，魏晋风尚，吴带当风，笑靥永恒。及于宋元明清，薪火不泯，当今之世，和光同尘,有识之士,尽心力,展才华,再造精神空间龙门博物馆应运而生。圆融无碍,圆满无尽。值《龙门博物馆建筑艺术》出版及龙门博物馆建筑艺术展开展之际,谨缀数语,为龙门庆、为祖国庆、为世界庆!

龙门博物馆馆长
王迪

Introduction

Longmen Museum is situated on the land of China by divine work as well as by the foremost ancestors of our nation. It carries a long history of Chinese civilization far and wide. The bird of Jiugao Mountain sings day after day, while Luhun Brook ripples gently. Between the two mountains, majesty finds its expression. Integration of man and nature makes available this very masterpiece. Buddhism is introduced eastward to Luoyang, light of Buddha shining on ten thousand statues. The clothes look as if they were just washed in water, which style being of Wei and Jin dynasties.with sleeves and belts seem to dance to and fro, they each smiling an eternal smile. Even up to Ming and Qing dynasties, Longmen Grottoes retains its original charm. Today, however, joint efforts are made to build this museum for the purpose of opening up a spiritual space to re-present its original harmony and perfection. Upon the publication of the Art of Construction of Longmen Museum as well as the display of Art of Construction of Longmen Museum, I'd like to say a few words to extend my congratulations!

Longmen Museum
Manager Wang

序言

　　相信并不是我一个人，把前言，本应最先完成的工作放到最后来作吧。现在很后悔，因为好像把力气都使完了，这本应该最用心的文字倒像是应付一样。不过也好，轻松一些吧。之前把精力都放在理念阐述那部分了，初衷是为了在最短的时间里配合排版的需要。本来就对写东西没有信心，写起来后发觉还是低估了难度。建筑师从来就是很忙的，至少是自我感觉良好的建筑师。当你不得不放下手边的工作，思索要表达什么时，才发现平时是多么的缺乏整理。而好像是老天冥冥中的安排，一个多星期以来连续的出差，给了我很多在路上的时间，才勉为其难地写完了理念阐述的部分。现在的我，也是坐在老挝首都万象市的街角咖啡，身边躺着两只慵懒的猫，偶尔抬头看我一眼，又继续闲适地靠着伙伴睡去，好像在说："不要执著，回头是岸"。看来传统佛国的小生灵们都是很觉悟的。"能闲世人之所忙者，方能忙世人之所闲"。

　　发现写作其实是个好东西，你会从一开始的疲于应付，慢慢地融入文字的世界中，梳理自己的心绪。慢慢发现还有另一个自己。正像法国思想家布朗肖的话："所谓写作，就是要发现异己，把思想里面那个不认识的自己发掘出来，写作永远是遭遇一个相异的人。"

　　这次由于时间压力过大，一直是赶着写，但也觉得收获颇丰了，下回如果闲情一些，一定是件快乐的事情吧。随性一些，更符合我的本性，也许这样能表达的更率真些。

　　这本书是应龙门石窟王迪馆长的要求制作的，虽然我觉得早了一些，6年多都等了，还差半年就真的开馆了，到时候再出不是更好？然而他说到时再出下册，重点是完工后的实景，我想也对，就同意了。王馆长是个性情中人，这首先基于他的职业。他是个收藏家，一个至少是我见过的最大的收藏家，虽然我也没见过几个。然而他收藏的文物珍品品质之高，数量之多，品种之丰富是足以让我这个门外汉咂舌的。尤其是其珍藏的洛阳永宁寺出土的泥塑，其令人震撼的生动，令我看到后有想落泪的冲动。塑像都很小，也大多残损，而存在的部分已充分地表现出佛教艺术中那份纯洁的虔诚。更是在混迹于泥土千年之后，逝去的色彩反而绽放出夺目的艺术光辉。王馆长时常给我展示他的藏品，那样子甚至有些像炫耀自己玩具的孩子，充满了无邪与真诚。而我偏偏不是个善于口头表达的人，虽然心中无限敬仰，嘴里却说不出什么赞扬的话，想必他心中在暗骂我不懂欣赏吧。王馆长中等身材，相貌朴实，不修边幅，但胸怀锦绣，机敏睿智，眼光独到，而且为人亲和，广

Foreword

I believe that I'm not alone to complete a foreword, which ought to be finished first, as a final step. Now I really regret for I appear to be exhausted, so it seems that I just deal with this text casually though it should be handled with greatest care. Anyway, it's better to get relaxed. I originally spent most of my efforts on the section of concept interpretation, with intention to meet the requirements for typesetting and layout in the shortest time. Frankly, I'm not very confident about my writing. After I commenced it, I felt that its difficulty was really underestimated. Arechitects are always busy, at least,the ones who feel good about themselves. When you have to put aside your work to contemplate what to express, you'll discover you almost never sort out your thoughts. Fortunately, more than one week's business travel came by destiny gave me more time on the way and enabled me to barely write out the section of concept interpretation. Today I'm sitting in Café Corner located in Vientiane, the capital of Laos. Beside me are two indolent cats, which just look at me occasionally and then continue to sleep cozily leaning against each other. They seem to say, "Do not be so clinging; repent and you'll be saved." Seemingly the little creatures in a traditional Buddhist country are also enlightened. "Only those who take leisurely what other people are busy about, can be busy about what other people take leisurely."

I found writing is a good thing; you would be slowly integrated into a world of words to groom your feelings after you could hardly deal with it from the very beginning. Gradually you'll find there is another yourself. Just as French thinker Blanchot said, "The so-called writing is to discover a dissident, namely, digging out that unknown yourself in your mind. Writing will always encounter a different person."

Given the super tight schedule, I'm too pushed to write it but also benefit greatly from the process. Maybe I'll be happy next time I am engaged in it with a more leisurely and carefree mood. Being myself better suits my very nature. This may help me portray a real self.

This book is produced complying with a request from Wang Di, curator of Longmen Grottoes. I think it is a little soon compared with over 6 years we have waited for. Will it be better to publish a book when the museum opens to the public in half a year? However, he says he prefers to publish Volume II subsequently to highlight the actual scenery upon completion. I think it right and thus agree with him. Curator Wang is a simple person with true feelings, which is first bound to his profession. He's a collector, at least the greatest collector I have ever seen, though I haven't met too many. However, the treasure of cultural relics he collects features highest quality, is countless and has various categories, which is amazing to a layman like me. In particular, the clay sculptures excavated from Yongning Temple of Luoyang present incredible vividness and almost move me to tears. Most of sculptures are small and broken, while the existing parts are still be able to reflect the purest godliness in Buddhism Art. Through a thousand year's mixing with clay, the

而改变了功能，设计成一半办公一半机房的综合建筑。结果投标时技术分为零。然而幸运的是我们有一个好的理念。更幸运的是评委主席是郑州的著名设计人肖艳辉老师。我们从不认识，而他不但力排众议评我们是第一，还介绍我们认识了王迪馆长，推荐我们设计龙门博物馆。而在此之前，他们已经运作龙门博物馆三年多了，并有了方案，还是某位大师的作品，而肖老师是不迷信所谓大师的。现在谈谈肖老师吧。肖老师与王迪一样，浑身散发着质朴与睿智，身材清瘦些。但属于体形不大但容量惊人型的。他从不张扬，沉静时是严肃而深沉的，但是大脑内的判断与运转是如此的高速与高效，并且不是在一个时空与领域内，他是在纵贯古今，并横跨哲学与艺术及科技，在人性与灵性中权衡。而当他发言时，一改深沉与整肃，变得生动诙谐，引古论今，语出惊人。他知识面极广，且有选择性地吸收，这一方面取决于他的职业，一方面取决于他的智慧。他是位设计师，虽然是学工业设计出身，但现在主业是室内设计，而他同时也是位收藏家，用他的话说主业还是逛古玩城。所以他会把对收藏的理解，对艺术品的鉴赏用在设计上。他非常关注理念，关注文化，反对装饰，反对矫揉造作，并且不轻易妥协。在谈论设计的时候，肖老师非常善于用比喻来说明问题，每次引得听众心驰神往又回味无穷。而最令我折服的还是肖老师的智慧。他不信佛，但他很禅定，也很觉悟。他是懂得平衡的人，无论哪一方面。他每天只上午上班，二三个小时安排好工作就去干别的了。中午雷打不动两个小时桥牌，下午回家看书或处理私事，晚上九点关手机。他身边的人都知道他的习惯，决不会在他打桥牌时打电话给他。他只做设计，从不干工程，用他的话说这钱他不挣，太累，并且会让自己迷失设计的方向。他收藏的文物都是他喜欢并符合他审美标准的。他不喜欢的，多好都不要，但碰到真喜欢的，可以拿出以往的东西去换。他不算很富足，他的藏品也不是很多（相对王迪），但他的内心很富足。他谈论文物时和王迪是不一样的，他也眉飞色舞，但不会忘形。讲的也都是细节与整体之间通融的艺术表现力。他对造型更敏感，对美有独到的见解。王迪是个收藏大家了，藏品多过肖老师不知道有多少倍，但他也经常拿这新收来的东西给肖老师看，并渴望得到肖老师的

上: 肖老师; 右: 丹尼尔

Above: Mr. Xiao; Right:
Daniel Schwabe

...lapsing colors emit glittery art lights. Curator Wang often shows me his collections just like a boy showing off his own toys, filled with innocence and sincerity. Unfortunately, I'm a little weak in expressing myself. Though I respect him with all my heart, few compliments can be heard from me. I guess he must be scolding me in his mind for not understanding him and appreciating his collections. Curator Wang is of medium height and plain looking. He doesn't value his appearance, but he is knowledgeable, easy-going, agile, wise and farsighted. He has deep insight into everything and makes friends with many people developing good relationship with everyone. He believes in Buddhism very sincerely. He loves Longmen and firmly believes that his career is to serve Longmen and benefit Longmen. We have been engaged in design of museums for nearly 7 years, while he started planning it 10 years ago. Just as a saying goes, "Persistence is the key to success." But he is also as with other common people and has various emotions and desires. Thus nobody can imagine the difficulty he experienced in these 10 years. However, he doesn't care about his personal feelings. With the strong ideal and belief in his heart, he finally came here today, step by step. If he has something exceptional than others, that's his penetrating judgment, striking art appreciation competence, and sensibility and easiness as an artist. The latter also makes him less organized and more amicable, which is very similar to me. Just as an old saying goes, "People of a mind fall into the same group." Curator Wang's complexion appears to be enchanted with Buddhism, especially his modest but profound eyes. The arcs in his canthi are also special, having the legendary Buddha verve in statue art. His natural smile can infect everyone close to him, like happy Maitreya Buddha but emitting a humanistic glow.

We got to know each other in September 2004 when my partner Daniel Schwabe and I just returned to China and got involved in the bidding for a project in Zhengzhou, which was another legendary story. That project was about Hub Building of Eastern Zhengzhou New District for China Mobile Zhengzhou Branch. As another saying goes, "Those who know nothing fear nothing." Given the uniqueness and significance of the project's location, our design plan didn't include an equipment building (hub building is a mobile program-controlled equipment building) as requested by the design document, but changed the function and designed a comprehensive building with one half as office area and the other as equipment room. As expected, we got a technical score of 0 during the bidding process. Fortunately, we presented a good concept and, more fortunately, Mr. Xiao Yanhui, a renowned designer in Zhengshou, acted as chairman of Evaluation Committee at that time. We never heard of him before, but he prevailed over all dissenting views and finally chose our design works. Then he referred us to curator Wang Di and recommended us to design Longmen Museum, which was run by them for more than 3 years prior to this. They even got a plan from a famous design master. But Mr. Xiao...

...never blindly worshiped a so-called master. Now let's talk about Mr. Xiao. As with Wang Di, Mr. Xiao emits modesty and wisdom from all over his thin body. He is average sized but always energetic. He never tries to attract attention from others, but his calmness is serious and deeply immersed in his world. His brain always judges and runs quickly and effectively across various times and spaces. He reflects over both ancient and modern times, covers philosophy, art and science, and weighs between humanity and spiritualism. When he delivers a speech, he will thoroughly change his deep seriousness and turn to dynamics and humor, with stunning words to capture everyone's attention. He's knowledgeable and can absorb everything at his own discretion, which is attributable to his procession as well as his wisdom. He's an interior designer and now mainly engaged in interior design though he studied industrial design in the university. At the same time, he's also a collector and his main business is to wander about antique stores everywhere according to him. So he tends to use his understandings of collections and appreciation of artworks in the design

人可，有时认真得像个学生一样。他经常到郑州找肖老师，从文物收藏到人生感悟，大到事业方向，小到家长里短，与肖老师聊得很多。这次博物馆的设计，就是王迪拜托他操心的。而肖老师不爱操太多心，也有时间就会和朋友自驾游出去一个月，每个周末一定在古玩城，到北京来一下火车就直奔潘家园。能认识他真是我们的幸运。感谢中国多动。

丹尼尔是我的合伙人，也是我最好的朋友，我们两个2004年一起回到中国，直到2009年他带着他中国的妻子回德国。他是我在德留学时第一个作业的合作人，也马上成为了最好的朋友，当时还有福客（Volker Kunst）。说来有趣，我们两个回国时，福客已经在中国半年了，是因为我介绍他去中国当老师，就在郑州。我们回国也是福客的建议，本想回来看看形势，结果就一直待到现在。龙门项目是我与丹尼尔一起开始的，而这时福客却已回去了。我们回国之前，在德国注册了我们的事务所Synarchitects,为了令中国人好念，我们干脆就简化为SYN建筑师事务所了。我们两个在中国的创业之路写起来就没有止境了，好在龙门博物馆设计开始时，我们回来才不久，直到丹尼尔回德国去还没有结束。实际上他走时才开始挖基坑，所以龙门博物馆设计的最艰难时期，是他陪我度过的。

丹尼尔是个很有趣的人，用中国人话讲，他不像个传统思维中叛版的德国人。他十分理想主义，比我还要严重得多。他总是对未来充满希望，对自己十分自信，对中国充满好奇。他个子不高，甚至还没我高，这又是一个他有别于一般德国人的地方。他也十分随和，爱开玩笑，经常和我搞点恶作剧。他十分容易兴奋，自然也容易沮丧，他嘴里常说的话是"Geil"和"Hammer"，"Geil"是太棒了的意思，北京话就是"牛"了。"Hammer"是锤子，也是很棒的意思，用现在的流行话就是"给力"了。从他们口头语大家也可以看出他是多么不吝惜自己的愉快心情了，也因此令大家都很喜爱他。丹尼尔还极聪明，总会有奇思妙想，与他一起工作是件十分愉快的事。现在，当我想起他时，脑子里总是他在对面电脑后跟我讲他的好想法时的样子，两眼放着光，一副要把我融化的架势。很可惜他现在不在这里。写到这里，大家可能会觉得他激动有余，定力不足吧，其实他工作时是十分专注的。当他进入工作状态时，会几个小时不同我讲话，他会在细节上毫不放松，也很难对现在的工作满意，就算满意也会在第二天推翻。同时他还很谦虚，会认真与你讨论，并虚心接受不同意见，当然也会有争执，但他也不会固执，无论谁说服谁他都会对结果满意，完了还是那句"Geil!"现在写这些时，真的很想念他，我们当然会经常通话，但是很希望能再次一起工作。他现在回柏林经营Synarchitects德国分部了。祝愿他一切顺利。

龙门博物馆的设计历程中，以上几个人是最重要的参与者。王迪作为业主从来没有怀疑过他的选择，并始终对我们支持与鼓励。即使是最艰难的时候，他都没有想过放弃，也没有对我们失望，因为他喜欢我们的理念，并坚定的站在我们这边，无论哪个领导或专家提出质疑，他都第一个站出来反驳，好像这是他的设计的一样。他还经常把方案拿出来给别人看，就像他珍藏的文物一样，得到肯定后一定会得意的打电话给我，一副心满意足的口气。多么可爱的一个人！目前是王先

...e emphasizes concepts and cultures but opposes decoration and artificial manners. He never wants to compromise. Mr. Xiao is skillful in using metaphors to interpret the questions when discussing the design, attracting audiences greatly and leaving a lasting and pleasant impression on them each time. However, what I admire most is still his wisdom. He doesn't believe in Buddhism, but appears to have many meditative experiences and have been enlightened. In every aspect, he really understands how to live a well balanced life. Every day he goes to work in the morning only, spends 2 or 3 hours in arrangements and then go for something else. Playing bridge for 2 hours at noon is never altered under any circumstances. And in the afternoon, he just stays at home reading or dealing with his private affairs. Finally his mobile phone is turned off at 9:00 PM. People around him all know of his habits and never bother him during his bridge game. The assignments he undertakes are limited to design projects only but not engineering projects. As he said, he never wants to earn profits from engineering ones because they are too boresome and likely to make him lost his way in the design. The cultural relics he collects are his favorite and meet his aesthetic standards. But he never collects what he dislikes, no matter how the cultural relic is valuable. For anything he truly likes, he can give his previous collections in exchange. Though he isn't very wealthy and doesn't have too many collections (compared to Wang Di), he's really fruitful in his innermost heart. His comments on cultural relics are also different from Wang Di, e.g. enraptured but never be lost in exhilaration. What he talks about also involves the harmonious artistic expression between details and the entire works. He's extremely sensitive to sculptures and holds unique views about the aesthetics. Wang Di deserves the title of great collector and his collections are several times Mr. Xiao's. But he also brings cultural relics newly acquired by him to Mr. Xiao and aspired to be recognized by him. Sometimes his serious attitude makes you mistake him for a pupil. He often goes to Zhengzhou and visits Mr. Xiao. They discuss almost every topic, from collection of cultural relics, reflections on life, career development and family affairs. The design of the museum was also requested by him. Mr. Xiao doesn't want to be bothered too much; he usually travels with his friends by driving cars in person for one month in his spare time. You can always find him at antique stores on weekends and the first stop after he gets off the train in Beijing must be Pan Jiayuan Market. So it's our great honor to know him. Thank China Mobile!

Daniel is my partner and also my best friend. We returned to China in 2004, but in 2009 he went back to Germany along with his Chinese wife. He was the cooperator of my first homework when I studied in Germany, and then soon became my best friend. The other friend I met in Germany is Volker Kunst. Interestingly, he returned to China half a year earlier than us for I introduced him to work as a teacher in Zhengzhou. Following his advice, we originally wanted to experience domestic circumstance, but actually stay in China until now. I commenced Longmen project with Daniel after Volker Kunst returned to Germany. Prior to our departure from Germany, we registered our firm Synarchitects. To make it easy to read by Chinese people, we simply abbreviated it to SYN Architects. When it comes to its startup and growth, it's really a long story. We just returned to China when the design of Longmen Museum started. And we didn't complete it before Daniel returned to Germany. Actually, the foundation pit was just excavated when he left, so it's him who spent the most difficult situation with me in the design of Longmen Museum.

Daniel is very funny or, according to Chinese people, he isn't like a traditional German with inflexible thinking. He's more idealistic than me. He's always optimistic about the future, confident about himself and curious about China. He isn't very tall and even isn't taller than me, which also distinguishes himself from other common Germans. He's also very easy-going, likes playing jokes and often plays tricks on me. He easily becomes excited and vice versa (depressed). The words that occur most often out of his mouth are "Geil" and "Hammer". The former means Great, which is equivalent of Niu (awesome) in Beijing dialect; while "Hammer" also refers to Well Done. In this regard, you'll feel how he's always ready to bring his pleasure to others with no stint and, of course, he's really favored by everyone. Daniel is ingenious and often presents his unusual and wonderful thinking. Working with him was really a pleasant experience. Now his appearance still appears vividly in my mind every time I recall him. For example, he used to discuss his good ideas with me from behind his computer and tried to convince me, with eyes glinting and staring at me. It's a pity that he isn't here now. Seen from what I described above, maybe you'll say that he's over energetic but lacks concentration. Actually he's always absorbed in everything he worked on. Once he started his work, he wouldn't even say a word to me for several hours. He's strict in every detail and hardly satisfied with his current work. Even though he's temporarily content with his work, he would overthrow his conclusions next day. Meanwhile, he's very modest, could discuss with you carefully and listen to different opinions. Sometimes we might disagree with each other, but he's never stubborn and always ready to accept results no matter who convinced who. And at the end of discussions he would present his "Geil" as usual. Now I really miss him very much when I'm writing here. Though we often communicate with each other by telephone, I really hope I can work with him again. Now he's operating Synarchitects (Germany) in Berlin. Hope everything goes well with him.

紧张的时候，该项目得到洛阳市政府的充分重视，定为河南省重点工程。而且新的一届洛阳牡丹花会就要开幕了，他正急着在没完工的建筑里筹备两个展览，其中一个就是建筑方案展示。这也是这本书急着出生的原因，希望不会因早产而智障吧。肖老师答应写一篇文章加入本书，应我的要求，写一写他从文物收藏中获得的艺术感悟，很期待他第一次落笔的成果，相信一定同他谈话时一样语出惊人。肖老师在方案之初没给我们明确的期望值，只是希望我们发挥才能，并表达了对博物馆的重视。他是不想局限我们的思路，给予我们足够的空间去琢磨，而且还对一旁火急火燎的王馆长说不要太着急，让他慢慢想，好东西从来都不容易得到。当我们第一次把概念拿出来时，肖老师沉静的听完我的陈述，稍微沉吟后果断地否定了我们的结论，并提出一个问题："为什么这个建筑是放在龙门，而不是放在欧洲或任何一个国家呢？"当时我们什么都答不上来。然后他又肯定了我们理念的前半部分，就是"信仰的力量"与"可视世界与理想世界的过渡空间"。只是这不是我们要的最佳答案，关乎龙门与佛教主题灵魂的东西还没有产生。他提示我们关于写意山水中的无以名状的空灵境界以及中国文化的"大象无形"。关于他提的那个问题，其实是说建筑的地域性与文化性的缺失。

我们很沮丧的回了北京，尤其是丹尼尔。他满腔热情与满怀自信，他觉得这个方案很好。这件事带给我们良久的思考。我又开始研读佛教的哲学，这一读就是近两个月。这两个月是我们建筑职业生涯可能是最重要的的一段时间了。正如我在理念阐述中所说，我们是在牛角尖中看问题的，只不过是从这个牛角尖跳进另一个大一些的牛角尖而已。我至今认为自己没有读懂佛教，但在那段时间的研读后感到"性空"的佛教世界就在身旁了。我们最大的敌人就是我们自己，是我们的"执"与"迷"在作祟。当我们最终开始动手时，丹尼尔也已经在电脑里做了很久的尝试了。他是一个电脑的依赖者，在我们工作的过程中，发现往往过于直观的东西太早地出现在眼睛里时，它是没有足够的表现空间的，往往不太明确的形，例如草图反而带有更多想象与隐喻的空间，这一点他自己也承认。于是我们两条路并行，我自始至终在画草图，他在电脑前验证与试验。记得是在一个酒吧里，我们有了"内空"的雏形。并由此设想了流线与功能排布。然而还是觉得少了些什么，是悸动。我们少的就是对情感的呼唤，一个不用解释就能打动人的东西，一种"美"的功能。

这时，投标开始了，我们有了竞争对手。王馆长仍然把肖老师当做评委主席，因为他知道肖老师是公正与严明的裁判。他不会为任何虚伪与伪饰所迷惑，他当然是对的。经过上一次的被否定，我们这次小心了很多，精心地为汇报做着准备。丹尼尔显得更是紧迫，可能是我比他大几岁的缘故吧，还稍微好些。我们很快将我们在酒吧得出的空间做出了模型，继续对其功能性及空间影响力做进一步的研究。这时丹尼尔在一次偶然中看到电脑中一个镜像的倒影图像，我们的眼睛都为之一亮，马上果断地得出答案——"就是这个！"。看来任何的努力都不是白费，当时我在速写本上写了一句话，"所谓的必然性，就是将你所有的可能性发挥到极致。"可惜那个速写本已经遗失了，里面记录了大量的草图历程，应该也会是这本书中重要的素材了。从这段经历，我们

These persons are the most important participants in the design process of Longmen Museum. Wang Di, as owner of the Museum, never suspects his choice and has been supporting and encouraging us all the way. Even at the most difficult moments, he never thinks of abandoning or feels disappointed with us because he likes our concept and firmly stands on our side. No matter which leader or expert questions it, he'll be the first to defend it as if it was his own design works. He often presents the plan to others as a cultural relic he collects. Receiving a positive feedback, he'll surely call me with a completely satisfied tone. What a lovely guy! Now it's the busiest moment for the working site. The project is valued highly by Luoyang Municipal Government and was evaluated as one of major projects in Henan Province. Moreover, the new session of Luoyang Peony Fair is forthcoming and he's busy preparing two exhibitions at the avenue to be completed, one of which is to showcase the building plan. This is also the reason why we rush the publication of the book and hope it won't have too many defects due to early production. Mr. Xiao promises that he will write an article as an addition to the book. Complying with my request, he intends to describe his artistic comprehension from collections of cultural relics. I'm really looking forward to his first writing results and hope they will be as amazing as his talking. At the beginning of the plan, Mr. Xiao didn't set clear expectations for us but just hope that we can bring into full play our potential given the significance of the museum. He didn't impose any restrictions on our thinking and left enough room for us to ponder. Seeing curator Wang who was eager to get the plan, he even persuaded him to calm down since the best things are always difficult to seek. After we presented the concept for the first time, Mr. Xiao thought it over for a little while and then decisively denied our conclusion after listening to my presentation silently. He put forward a question, "Why is this building to be placed in Longmen, but not in Europe or any other country?" Neither of us could answer it. Then he also agreed to the first half of our concept, i.e. "Power of Belief" and "Transitional Space between Visible World and Idealistic World". But this wasn't the best answer wanted by him because we didn't generate something regarding the very essence of Longmen and Buddhism themes. He provides us a prompt on indescribable void and inspired state in landscape painting, and marvelous spectacles in unlimited forms and layouts. Actually, the question he mentioned referred to the shortage of the buildings' locality and culture.

We were so frustrated to return to Beijing, Daniel in particular, because he always believed it's a good plan, filled with passion and confidence. This caused us to ponder upon it in a long time and I also had to start reading Buddhist sutras, which lasted nearly 2 months and represented the most important time in our career as architects. As I said in concept interpretation, what we saw are insignificant or insoluble problems, perhaps we just jumped from this one to another bigger problem. So far I still don't think I understand Buddhism. However, I really felt a "empty state" Buddhism world beside me thorough a long-term reading. Our biggest enemy is ourselves and it's because we were so clinging and addicted that we were annoyed greatly. When we were determined to start finally, Daniel had attempted many times on the computer for he used to rely upon it. During our work, we discovered that the extremely intuitive things emerging in our eyes too early often lacked enough expression power, while a vague shape like a sketch would bring more imagination and metaphors to people. Daniel also recognized it. Thus we employed a dual-track method in which I had been painting sketches and he would verify and test them on the computer. I remembered that it's in a bar that we had a rudiment of "inner emptiness" and subsequently conceived a streamlined-functional layout. But we still felt something missing, yes, it was emotion. We failed to trigger emotions, i.e. something that can move people greatly without explanation—a function of aesthetics.

Then the bidding started and we encountered our competitors. Curator Wang still appointed Mr. Xiao as chairman of Evaluation Committee because he knows that he is an impartial and strict referee. Mr. Xiao's conclusion is absolutely right because he will never be deceived by any fakes or disguise. We were more careful and deliberately prepared for the presentation since we were already denied last time. Perhaps because I'm several years older than Daniel, he looked very anxious but I was calmer. We quickly set a model for the space we discovered in the bar and further studied its functionality and space influence. Just then Daniel occasionally saw an inverted image of a mirror image, and we were suddenly inspired and immediately gave the answer-"That's the case!" It seems that your efforts will never be in vain. I wrote such a sentence in my sketch book at that time, "The so-called inevitability is to bring into play your possibility to the fullest extent." It's a pity that I lost my sketch book afterwards, which recorded a multitude of sketches all the way and was an important material for this book. Viewed from this experience, we understood that something beautiful is never easy to obtain.

On the bidding day, we successfully reported our plan and then entered Bai's Garden (a landscape area developed based on Bai Juyi's tomb). There was no conversation between us because the result would inevitably determine our efforts for half a year. Thus we couldn't inhibit emotional fluctuations in our innermost heart. When we walked slowly and returned to the venue, we saw Mr. Xiao sitting beneath the tree leisurely from a distance. We weren't sure if we should say hello to him since the final results were to be announced

也明白美好的东西从来不容易，经验告诉我们，它之所以难是因为它什…
挂得。

投标那天，我们顺利地汇报了方案，因为要等待别家的汇报与投标
结果，我们两人走进"白园"（白居易的墓为基础的园林景区），没有什么
对话，因为这次的结果可能不可逆转地决定我们半年来的努力。我们难
以抑制内心的波动。当我们缓步按约定的时间步回会场时，老远看到肖
老师悠闲地坐在树荫下。因为还没有出结果，我们一时还不知该不该与他
打招呼，毕竟他是评委主席嘛。看到我们走过来，肖老师一脸轻松地说：
"这回行了，大问题解决了！"我们的心一下就归位了。评审结果是十个评
委九票对一票，我们中标了。奇怪的是，我当时真的没有狂喜的感觉，只
是松了一口气。之后的几年证明，中标也只是刚刚开始而已，艰辛之路还
在建设，路障正在调集，尽头还不知在哪里。

现在每当想起这段历史，最清晰的还是肖老师在会场门口的轻描淡
写的那句话。他当然对我们说过很多话，特别是在中标后。我想我还是把
一些东西留到博物馆建成后再说吧，让我们最后的努力变成现实，用美
好的空间换取肖老师与王馆长对我们一贯的信任吧。因为在最终看到她
之前，我的内心还是惴惴不安的。我怕的是，是否我又跳进了另一个大一
些的牛角尖。

2011年4月3日
于老挝万象市

Anyway, he acted as chairman of Evaluation Committee. But he looked relaxed and directly told us, "it's over now, we solved a big problem!" Then our hearts finally settled. The evaluation result was that nine of ten committee members voted for us, so we won the bidding. Surprisingly, I wasn't filled with joy at that time but just relaxed a little bit. The following years demonstrated that winning the bidding was only a start. We're still in a difficult environment and the barriers are assembling, but we don't know where the end of the road is.

Now I still clearly remembered what Mr. Xiao said casually at the entrance of the venue every time I recalled this history. Of course he gave us many instructions, especially after we won the bidding. Maybe it'll be better to say something after the museum is completed. Only through turning our efforts into realities can we deserve the consistent trust from Mr. Xiao and curator Wang with a magnificent space. Anyway, I'm still anxious and fearful in my innermost heart until I finally see it. I'm afraid that I've jumped into another insignificant or insoluble problem.

April 3,2011

Vientiane,Laos

Zou Yingxi

阳春随笔话设计

肖艳辉

　　三月的一日与王迪馆长的一通电话述聊后，突发兴起。王馆长驱车从洛阳赶来，又叫小邹从北京打飞的到郑州。三人就龙门博物馆建设的有关事宜相聚家中小院，在春天的暖阳下，忙里偷闲畅谈一天。王馆长深通华夏古代文化和艺术，小邹留学德国学设计，而我是喜爱收藏的设计人。相言东方与西方，远古与现代，纵横交错，颇有感念。今随笔杂谈信马由缰。

　　当今所言中式风格大多为明清式样，这在五千年的华夏文明史中如沧海一粟。因为逐鹿中原的过程中大多物化的东西已不复存在。今天所能见到的只能是古文物，这其中所传达出的穿越千年依然散发诱人魅力的艺术和精神之美才是真正的永恒。只有文化的才是经典的。这份厚重不是明清式样和风格所能代替的。只读历史也不能感受意象造型审美的中华文化。在玩收藏的过程中，仿佛能与不同的历史风貌对话。感知那份盛世繁华，那份颠沛磨难，那份超然世外，那份激情澎湃。形式不在，精神永存。真正的风格式样应该是一种内在精神气质的流露，而不是简单形式上的表现。能把中国文化的精髓融入设计师的血液中，那么不管嬉笑怒骂皆能出手成章。没有僵死的形式，只有灵动的精神。今天的中国正快步走向世界，中国文明的历史正是海纳百川的历史。中庸文化向来注重"度"的把握。而不是只强调"更高、更快、更强"地把人类送上不归路的单一思维模式。天人合一的大文化，是历经千锤百炼的结晶，让中华文明传承几千年而独立于世界。　从系统的高度看待和解决局部的问题，将单一的作品与天、地、人完美地和谐统一，永生相伴。　所以设计不只是美学问题，而是系统的管理问题，是集人类所有学科与智慧的完美统一。大作无形，大乐无声，大象无影。正是如此辩证的思维高度引导我辈设计出"看不出设计的设计"。浑然天成，妙手偶得。这不是技术的展示，是修为所得。

　　在龙门博物馆即将建成之际引发感怀。踏青逐日那天犹记，今随手写下：只有民族的才是世界的！有形无形妙在"中"！

An Essay on Design

Xiao Yanhui

Inspiration occurred to me a day this March after a word on the phone with Wang Di, the curator of Longmen Museum. Wang drove over from Luoyang, while Zou was asked to fly here to Zhengzhou from Beijing. It was a warm spring day. The three of us gathered in my courtyard, starting the discussion over the construction of Longmen Museum. Wang has a profound knowledge of our ancient Chinese culture and art, Zou has studied design in Germany, while I am a designer myself interested in collection. When it came to the topic concerning east and west, ancient and modern, I had quite a lot to share with them.

The Chinese style now talked about is mostly that of Ming and Qing dynasties, which is only a drop in the bucket compared with the 5,000-year Chinese civilization, for most of the concrete evidence is out of existence due to a fight among rivals for throne in Central Plains. What are available today are nothing but antiques which exhibit eternal beauty in artistic and spiritual charm that are still there through thousands of years. Only cultural things are classics. They have a deep historic meaning that cannot be replaced for the style of Ming and Qing dynasties.One cannot expect to perceive the real Chinese culture and its stylistic beauty only by reading some history. When collecting antiques, one feels as if he could converse with diverse historical scenes, feeling their prosperity, vicissitudes, aloofness, and passion. The form is no more, but the spirit is for ever. The real style should be something out of inner qualities rather than merely shown in form. If a designer can integrate the essence of the Chinese civilization into his own veins, then there will always be great works in his hand. No dead form, but lively spirit. Today's China is rapidly advancing forward along with the world. The history of Chinese civilization is of eclectic history; the eclectic culture tends to attach importance to the mastery of "measurements", rather than stress the single mode of "higher, swifter, and stronger" thinking, which may lead mankind to a one-way street. The broad culture of integration of man and nature is the product abstracted in history, inherited by Chinese civilization for thousands of years and unique in the world. From a system's point of view, we should always integrate a single work with the perfect harmony of man, heaven and nature, so as to consider and tackle problems of each part. Thus design is not merely a question of aesthetics, but also a question of management. It unifies all disciplines and wisdom of mankind perfectly. Great works need no form; great music needs no sound; great images need no shape. It is this dialectical thinking that has led us to producing "the design without design", which could only have been created by skillful hands to look like nature itself. It is not crafts, but cultivation.

Upon completion of the building of Longmen Museum, I'd like to present this essay, for which I first kept it as a diary and then wrote it as the present form. I believe: only those that are national belong to the world! Form and non-form are marvelous in "between"!

1973年5月—生于河南省郑州市

1987年—迁于北京

1993—1997年—就读于中央工艺美术学院（现清华美术学院）环境艺术设计系

1997—1998年—留校任教

1999年—荣获第二届中国室内设计大展金奖

2000年—赴德国留学，就读于柏林艺术大学建筑系

2004年—获得柏林艺术大学建筑学硕士学位,同年于柏林成立Synarchitects建筑师事务所(合伙人)

2004年—至今—在国内从事建筑实践活动,并任职于北京龙安华诚建筑设计有限公司, 任总建筑师

近期主要作品:

·河南省移动通信郑州分公司郑东新区枢纽楼

·广东省韶关市东堤路规划及建筑设计

·中央美术学院北京燕效新校区总体规划及建筑设计

·龙门石窟博物馆

·法国工业园项目

·宁波轻纺城

·宁波东方宜家花园项目

·宁波东方宜家广场项目

·宁波石契办公楼

·宁波江北奥特莱斯

·海南健康城项目

·怀柔制片人总部基地

·九龙湖赛车场

·宁波天波城项目

所获荣誉

2009年　韶关东提路改造经中国人居典范评审委员会评审获最佳设计方案金奖

2009年　中央美术学院燕郊校区经中国人居典范评审委员会评审获最佳设计方案金奖

2009年　东方宜家花园经中国人居典范评审委员会评审获最佳设计方案金奖

2010年　中国绿色低碳十佳建筑设计人物

2010年　亚洲绿色低碳（中国）示范豪宅（世纪宜家广场项目）

2010年　中国绿色低碳建筑设计创作大奖龙门石窟博物馆项目

May,1973 – Born in Zhengzhou, Henan Province

1987 Immigrated into Beijing

1993-1997   Studied on Environmental art design, Central Institute of Arts and Crafts (Now: Academy of Fine Arts of Tsinghua University)

1997-1998-Remained on the school staff

1999 Got the gold prize of Second term Indoor Architectural Design Exhibition

2000 Went to Germany and enrolled at the University of fine arts Berlin

2004 Diploma, Architectural Studies, University of fine arts Berlin,Same year,Founder & Partner, Synarchitects, Berlin

2006   Founder, World Chinese Union of Architects

2004-now   Master architect, Beijing Longanhuacheng Architectural Designing Co.,Ltd., China

Recent main projects:

·Central Building of Zhengzhou Branch of China Mobile Communications Corporation, Zhengdong New District, Zhengzhou, Henan

·Dongdi Road Planning and Architectural Design, Shaoguan, Guangdong

·Planning and Architectural Design Yanjiao New Campus of Central Academy of Fine Art, Beijing

·Longmen Museum   ·French Industrial Area

·Ningbo Light Textile Mall ·Eastern Ikea Garden, Ningbo

·Eastern Ikea Plaza, Ningbo ·Shiqi Office Building, Ningbo

·Jiangbei Outlets, Ningbo ·Hainan Healthy City

·Huairou Producers Headquarters, Beijing

·Jiulong Lake Autodrome ·Tianbo City, Ningbo

Awards & Honors:

2008 The project of Shanguan Dongdi Road Planning and Architectural Design was awarded the gold prize of the best design by Chinese Human Settlement Pattern Committee

2009 The project of Yanjiao New Campus of Central Academy of Fine Art was awarded the gold prize of the best design by Chinese Human Settlement Pattern Committee

2009 The project of Ningbo Eastern Ikea Garden was awarded the gold prize of the best design by Chinese Human Settlement Pattern Committee

2010 People of Top10 of Low Carbon Architectural Design in China

2010 The project of Ningbo Eastern Ikea Plaza was awarded the demonstration of luxury apartment about low Carbon in Asia.

2010 The project of Longmen Museum was awarded the Creative Award of Chinese Low Carbon Architecture Design

Daniel Schwabe

德国

建筑学，硕士

柏林艺术大学，硕士学位

工作单位 北京龙安华诚建筑设计有限公司

职务职称 设计师

从事规划设计的时间　　　6年

参加相关规划

项目经验和职责

1.郑州市郑东新区枢纽楼工程　　　　　方案负责人

2.北京京仪椿树整流器公司科研办公楼　　方案负责人

3.中央美术学院燕郊校区　　　　　　　设计师

4.龙门石窟博物馆及游客中心　　　　　设计师

5. 广东韶关东堤路规划及旧城改造工程　　主设计师

2004年　获得柏林艺术大学建筑学硕士学位

　　　　同年于柏林成立Synarchitects建筑师事务所(合伙人)

2004-2006年　在国内从事建筑实践活动；并任职于

　　　　　　北京龙安华诚建筑设计有限公司

2008年　参与设计的中央美术学院燕郊校区经中国人居典范评审委员会评
审，荣获最佳设计方案金奖。

Daniel Schwabe

Germany

Architecture, Diploma

University of fine arts Berlin

Company: Beijing Longanhuacheng Architectural Designing Co.,Ltd.

Position: Designer

Length of work: 6 years

Work experience:

1.Central Building of Zhengzhou Branch of China Mobile Communications
Corporation, Zhengdong New District, Zhengzhou, Henan

Scheme manager

2.Scientific Research Building of Beijing Chunshu Rectifier Co., Ltd.

Scheme manager

3.Yanjiao New Campus of Central Academy of Fine Art

Designer

4.Longmen Museum and the visitor center   Designer

5.Dongdi Road Planning and Architectural Design, Shaoguan, Guangdong

Designer

2004 Master, Architectural Studies, University of fine arts,Berlin

Same year Founder & Partner, Synarchitects, Berlin.

2004-2006 Architect, Beijing Longanhuacheng Architectural Designing
Co.,Ltd., China.

2008 Took part in the design of Yanjiao New Campus of Central Academy
of Fine Artand this project was awarded the gold prize of the best design
by Chinese Human Settlement Pattern Committee.

龙门博物馆设计方案评标会
Longmen Museum Architectural Design Scheme Bid & Evaluating Meeting

李振刚　　龙门石窟管理局局长
Li Zhengang
Chief
Longmen Grottoes Administration

王玮钰　　清华大学建筑学院教授
Wang Weiyu
Professor
School of Architecture, Tsinghua University

单德启　　清华大学建筑学院教授
Shan Deqi
Professor
School of Architecture, Tsinghua University

吴国力　　世界华人建筑协会秘书长
Wu Guoli
Secretary-General
World Association of Chinese Architects

可视世界与理想世界的过渡空间——龙门博物馆设计理念分析

The Space in the Transition from Visible World to Ideal World Analysis of Longmen Museum Design Principle

　　龙门博物馆的设计自开始思考至施工过程中的完善历时6年有余，其间无数的困难与无法预料的意外阻力不断出现，当然也有部分原因是由于我们与业主均经验不足导致项目受阻。回顾这6年的历程，曾经有过的多少次无奈、困惑以及从头再来，始终萦绕于心，并且这不仅仅是作为建筑师的我们，也是龙门博物馆王迪馆长感同身受的。幸运的是，建筑师与业主始终抱着共同的信念，互相支持与鼓励，以孩子般的理想主义与"玉不琢不成器"的决心应对所有的困难，这也就是所谓的"好事多磨"吧。

The design of Longmen Museum lasted over 6 years from conceiving to modifying the design during construction. Numberless difficulties and unexpected resistances kept emerging in the process, which was partly because of the insufficient experience of the owner and us. Over the six years, we have been resigned, confused and started afresh. As architects, we share the same feeling with Curator Wang Di. Fortunately, the architects and the owner shared a common belief, supported and encouraged each other, coped with all difficulties with childlike idealism and the resolution to be the best, just as the famous saying goes, "A good gain takes long pain".

从上至下: 施工照片
From above to below: the construction photograph

现在，博物馆已经封顶了，重新审视6年的经历，以及比较6年前与现在的我们，发现就如同博物馆从无到有的变化一样，我们自身也产生如此大的变化，建筑的过程也是我们完善自己与建构自己的过程。心中的蓝图快要接近现实的时候，写这段文字可以说是百感交集。而这里要说的是，虽然6年来龙门博物馆的建设经历了无数已经与可能的变化，然而有一样东西却始终如一的坚持着，不但丝毫没有改变，反而在经历了无数质疑与论证后更为清晰与坚定，这就是龙门博物馆的设计理念。它最初说服我们自己，随后征服了业主与评委，并在之后6年中不断演绎着自己的传奇，对话所有的关注，质疑，批评，否定，赞扬，建议，最终为所有人接受，并给予支持。

Now that the construction work is completed, we look back and find that we have been changing considerably over the past 6 years, just as what has happened to the Museum from a flat land to a splendid building. While designing and building the Museum, we were also improving and structuring ourselves. All emotions filled in my mind when I sat down and wrote these words. I wish to emphasize, though the many design revisions or proposed revisions were seen in the construction process, the design principle remained persistent without any change. Instead, it became even more clarified and confirmed after numerous questioning and studies. As a matter of fact, it is deemed as a legendary story which started with the principle persuading ourselves, the owner, and the evaluation committee, going through all concerns, doubts, criticisms, denials, praises and proposals and ending with being accepted and supported by all the people involved.

一.整理与准备
I. Summarization and Preparations

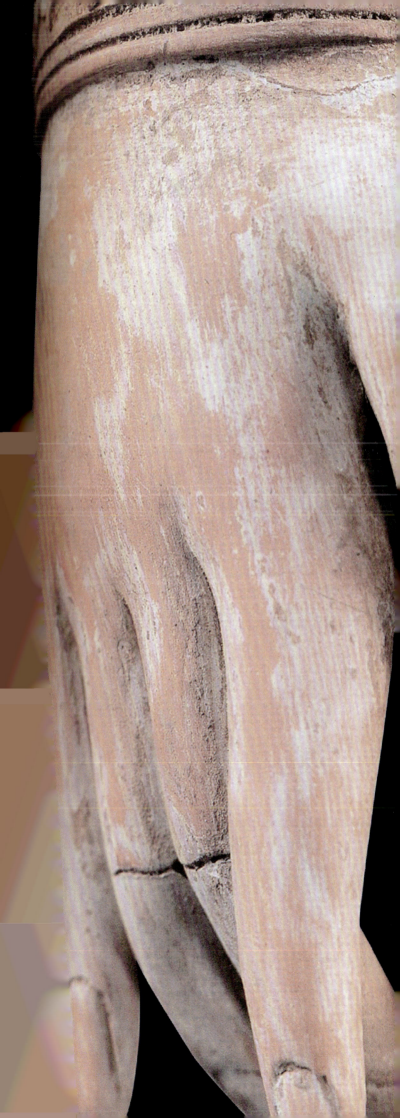

拥有1500年历史的龙门石窟，自北魏开凿以来，历经岁月的洗礼，朝代的更迭，静静地注视着中国文明的进程。她目睹了大唐的盛世，也俯视了无数的战乱与屈辱，不变的是她包容的胸怀，卢舍那大佛永恒的微笑，慈祥的关爱着中原大地的子民。如果说龙门石窟是中国第一次民族大融合（北魏）的产物，那么她也是历次民族融合的见证，如果说她以博大精深的佛教哲学敲开了中国的大门，那么她以普度众生的姿态敲开了中国人的心门。而真正令佛教融入中国人血液的，是她那伟大的圆融。这一圆融不但成为中国文化史的主导之一，也令佛教自身以包容的立场与道家及儒家思想相融合，不但产生了中国自己的佛教"禅宗"体系，也为"儒""释""道"三教各自寻求生存空间的同时，互相取长补短，共谋发展做出自身的贡献。事实上，禅宗是中国式的智慧对印度传统佛教单调教条的经院化逻辑的一种反动。

The Longmen Grottoes, after she was carved in the North Wei Period 1,500 years ago, has been witnessing the progress of the Chinese civilization in different eras ever since. She saw the flourishing period in Tang Dynasty, and numerous wars and humiliations suffered by the Chinese people. However, she has never changed her generous mind and the eternal smile on the face of the Grand Locama Buddha, showing gracious love for her people living on the Central Plains of China. The Longmen Grottoes was not only the result of the first ethnic fusion in China (that happened in North Wei Period), but also witnessed all the following national integrations. With the attitude of "universal salvation", Buddhism, a profound philosophy, came into China, but it was the interpenetration nature of the Buddhism that helped it get into the blood of the Chinese people. The interpenetration has not only become one of the dominant features of the Chinese culture, but also led to the integration of Buddhism into Taoism and Confucianism. Such integration has generated the Chinese Zen Buddhism, thus creating the separate development space for Confucianism, Buddhism and Taoism, which have learnt from one another and made their independent contributions. In fact, the Zen Buddhism, donated greatly by the Chinese wisdom, is the opposition against the traditional Indian Buddhist doctrines.

佛教在中国的发展除了自身哲学体系的发展外，也与中国5000年文化艺术相结合，产生思想与艺术的完美结晶。龙门石窟由联合国教科文组织选定为世界文化遗产是当之无愧的，她除了历经1500年的历史积淀外，还以2000多个洞窟，10万个造像的辉煌规模令世人惊叹。然而，龙门石窟历经多次战乱与外族侵略的洗劫，大部分造像遗失与惨遭破坏。龙门博物馆就是以龙门石窟佛教石刻艺术瑰宝的保护，研究，展示为目的而建设的。

另外，龙门博物馆作为佛教艺术专题博物馆还收藏了大量珍贵的陶瓷、青铜器、玉器、泥塑等佛教艺术珍品，其中大部分是王迪馆长多年的个人收藏以及全国多位收藏名家的藏品，当中尤以北魏时期洛阳永宁寺遗址出土的登峰造极的泥塑精品最为突出。当作为建筑师的我们看到这些艺术珍品时，深深为自己能够有可能为之建造博物馆而倍感自豪与荣幸。然而，什么样的建筑能够容纳这批美轮美奂的财富呢？其中的任意一件藏品，都绝不只是单纯的造型器物，而是在塑造伊始，创造者怀着对佛教崇高的虔诚，用心去雕琢的有灵魂有生命的艺术精品。其后又经过上千年血与火的岁月洗礼，使其原有的"形"更为模糊，更为含蓄，然而蕴含其中的纯洁与永恒的艺术价值反而更加深刻地凸现出来。我们设计的建筑要如何与这些艺术珍品对话呢？这首先取决于我们作设计时的态度，是否让自己与制作这些珍贵文物的艺匠们一样，怀着一颗虔诚的心去工作，是我们要首先做到的。然后，站在什么样的高度看待龙门博物馆，是第二个重点考虑的问题，她只是一座功能性的建筑吗？或者她除了是建筑还是什么别的吗？有必要令其承载更多的精神性的内容吗？如果是，又是什么精神呢？最后，才是形式问题，前两项的相加的结果是动机，而形式是结果。

但遗憾的是，艺术创作的动机与结果有着并非等比的关系。

While developing in China as a philosophy, Buddhism has also combined itself with the Chinese art and culture, producing the perfect ideological and artistic works. The Longmen Grottoes deserves her title as the World Cultural Heritage granted by the UNESCO because of her history of 1,500 years, over 2,000 grottoes and the admirable 100,000 carvings. However, most of the carvings were looted or damaged in the past wars and invasions.To protect, study and display the remaining Buddhist stone carvings, the Longmen Museum has been built.

As a Buddhist art museum, Longmen Museum also has a large number of Buddhism-related porcelain, bronze wares, jade wares and clay sculptures, most of them are the personal collections of Curator Wang Di and many other famous Chinese collectors. Among them, we are most impressed by the premier clay sculptures in the North Wei Period excavated in then Yongning Temple of Luoyang. As architects, we feel proud and honored to build a museum holding these art valuables. What kind of the architecture can hold such fascinating and valuable collections because each collection in it was created with the extreme piety for Buddhism, thus making it an art treasure with life and soul. In addition, these valuables surviving over 1,000 years have obscured their original "shapes" and highlighted their innocent and eternal artistic value. So how should we design the museum that can work in concert with these art treasures? The most important is that we should adopt the same pious attitude as taken by the craftsmen of these treasures when designing the museum. Second, should the Longmen Museum just fulfill its function as a building? To be specific, is it just a building holding the collections or is it something else that carries more spiritual contents? If yes, what kind of contents should it carry? Last but not least, what form should the architecture take, which is the result of the first two considerations?

Nevertheless, when making the art, the motive never produces the expected result.

1.佛头;3.菩萨头;5.菩萨头；2.双龙瓶；4.塔形舍利罐6.龙柄
小瓶（北魏以前）

1.Head of a Buddha; 3.Head of a Bodhisattva; 5.Head of a Bodhisattva.;
2. A White-Glazed Double-Dragon Bottle; 4.A Pagoda-Shaped Container
for Holy Relics; 6. Glass Vase with Dragon-shaped Handle

"形而上者谓之道，形而下者谓之器"，这句出自《易经》的话适用于一切造型艺术领域。从龙门石窟自身造像艺术伟大的圆融到永宁寺泥塑魁人的生动，都不是对形象单纯的模仿，而是融入了创造者内心的虔诚，与主观的意念。而建筑师面对场地时也应该把内心敞开，在吸收场地信息的同时，审视对象的精神潜质，挖掘其背后的思想内涵，至此，探索出对象的气质。正如玉石家们面对一块璞玉时，探究其内在的品质，并想象其最终的温润与深邃。而好玉如果在糟糕的匠人手中，最终还是劫数难逃的，被去掉的是朴实无华的美，刻画上的却是对美的玷污。所以主观的意愿或动机是基于对象的气质而发，并以"形而上"的"道"为目的，怀着惴惴之心，努力建构内心的美好图像，而且随着对对象了解的加深不断重构。艺术创造的过程就是内心不断重构，境界不断提升，对结果无限憧憬的过程。艺术家的境界高下，也就是其对结果广度与深度的把握的高下，并通过灵性有效的控制其重构与权衡对象的周期。从这个意义上说，也就是艺术创造没有止境的远原因，也是动机与结果非等比关系的原因。而灵性是游离于两者之间，并决定其关系的要素。例如舞者，我们总是很关心他们怎么跳，却从来不关心他们为什么跳，艺术不是应该从这最基本的动机问题开始吗？跳的动机又有其高下，而舞者是有决断力与行动力的代表，同时又是有灵性的人，人又有深浅与轻重之分，动机的高下，与人的灵性与否，再加上行动力，形成三重障碍决定了结果的不同。没有动机，或坏的动机，灵性与行动力产生的结果只能是零或是负数。没有灵性，动机与行动力的结果只能流于平庸。没有行动力，就只能"虽不能至，心向往之"了。建筑设计又何尝不是如此呢？

According to the Book of Changes (also called I Ching), "The things above the specific forms are called Tao while those below the forms are the utensils", which is applicable for all formative arts. The stone carvings in the Longmen Grottoes and the clay sculptures in the Yongning Temple are not the plain copying of images. Instead, they have been instilled with the piety and conceptions of the craftsmen. When facing the site, the architects should open their mind, absorb the emitted information, review the spiritual contents of the art treasures, explore the connotations in them, and seek the temperament of the collections, just as the jade sculptors would do in front of the crude jade. If a piece of good jade falls into the hand of a bad craftsman, it will inevitably end in a misery fate with its natural beauty eliminated or even stained. As a result, we can conclude that the subjective motive or vision is triggered by the temperament of the object to carefully construct the image in the mind in the effort to create the "Tao above the form". And as the understanding of the object goes deeper, the image keeps being upgraded. The artists differ from each other because of their different capabilities to grasp the extent and depth of their works, and to smartly and effectively control the restructuring and balancing of their objects. From this, we can understand why there is never the best artistic creation and why the motive never produces the expected results. Indeed, the so-called ingenuity is the key element lying between the motive and result and determining the interrelations between them. For instance, we only show keen interest in how the dancers perform but never ask why they should dance. Is the elementary motive the point where the art starts with? The dancers, who have their own determination and action power, are also the smart people. So the result depends on the three factors, namely, the determination, the power of action and the smartness. Without a motive, or without a good-willed motive, the smartness and action power will produce no result or negative result. Without the smartness, the motive and power of action will only generate the mediocre work. And without the power of action, we can only "cherish the result in our dreams". And such is the case with the architecture design, isn't it?

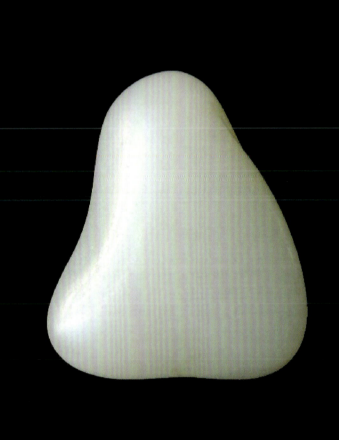

龙门博物馆的设计当然首先要解决动机问题。动机分三部分，如前文所述，第一是虔诚的用心，面对叹为观止的龙门石窟以及王馆长精美的藏品，我们岂止是虔诚，简直是敬畏。面对佛教哲学的博大，我们为自己的无知而羞愧。面对龙门石窟研究所的官员及王迪馆长对博物馆工作的认真与热情，我们由衷的敬佩与折服。只有用心地研读佛教的精义，学习龙门石窟的历史，吸收场地的信息，进而探究博物馆的气质。第二是以什么样的高度来看待我们设计的工作与对象，这一点不妨首先用龙门石窟研究所及管理局的领导的话来说明，他在投标前的工作动员中用"四个一体"来形容博物馆的意义。

1. 与龙门石窟景区环境融为一体

2. 与洛阳盛世的气势融为一体

3. 与佛教文化、理念、思想融为一体

4. 与龙门石窟景区的整体功能融为一体

这"四个一体"的出现，就决定了我们所要完成的设计任务已绝不仅仅是一座功能建筑那么简单了，她还承载着佛教哲学与艺术的精神。同时她的出现也是龙门石窟历史上的里程碑，是洛阳的新亮点。洛阳文博专家们还提出，博物馆对于洛阳的影响，要提到石窟对洛阳的影响高度。在中国经济开放与腾飞的大环境下，建筑要有大唐的胸怀与对话世界的语言，让世界人民通过她了解中国与中国的佛教文化。洛阳的领导与专家满怀的期望及传达出的概念无论多么宏大，多么高不可攀，甚至到了令人望而生畏的地步，但其实又那么质朴与合理，因为龙门石窟的历史与艺术价值的确配得上如此高的要求。而博物馆的位置，就在距龙门石窟北入口500米的位置上，处于石窟景区核心区内。景区每年接待近二百万游客，旅游高峰时，日客流量达7670人次，其中当然包括大量境外游客。而建筑物的出现，哪怕只是一幢一万平米的建筑，面对1500年悠久历史的龙门石窟时却是那么貌似渺小，实则影响力强大的物体。如此来看，如何处理这"四个一体"与建筑物实际对龙门产生巨大影响的现实呢？或者如何做才能令这种影响是积极与正确的呢？建筑物对龙门石窟的影响最终还是由形态、体量、材料、气质来体现的，如何来判断这几个要素的性质，以适应以上问题呢？这样就产生了动机的第三部分："建筑设计理念"。

In designing the Longmen Museum, the first issue is about the motive, which has three parts. As mentioned above, the first part of this issue is the piety. We have shown even more than piety towards the admirable Longmen Grottoes and the classics collections of Curator Wang. As a matter of fact, we show reverence for them. In front of the profound and extensive Buddhism, we feel ashamed of our ignorance. And we admire the Longmen Grottoes Institute officials and Curator Wang Di for their dedication to the work and studies. We fully understand it is imperative for us to learn the essence of Buddhism, probe into the history of Longmen Grottoes, absorb her information and explore the temperament of the Museum. The second part is about the perspective of viewing our design target. We believe the comments made by the Director of the Longmen Grottoes Institute and the Administration can best explain this part. At the meeting that announced the bidding, he summarized the meaning of the Museum into the following "4 conformities"

1.Conformity into the surrounding environment;

2.Conformity into the prime time of Luoyang;

3.Conformity into the Buddhism culture, ideology and thought; and

4.Conformity into the overall picture of the Longmen Grottoes scenic area.

Such requirements as the "4 conformities" have determined that our task was not just to design a functional architecture. Instead, it should carry on the Buddhism philosophy and art. At the same time, the Museum should be a milestone in the history of Longmen Grottoes and the new spotlight of Luoyang. The heritage and museum specialists of Luoyang pointed out that, the Museum would have the same influence on Luoyang as the Longmen Grottoes would have on the city. When the Chinese economy is rising robustly, the Museum should have the broad mind featured by the flourishing Tang Dynasty and the boldness to interact with the world people, who can understand China and the Chinese Buddhism culture by visiting the Museum. No matter what great expectations are voiced by the Luoyang municipal leaders and experts of the Museum, they seem rational and plain because the historical and artistic values of the Longmen Grottoes are always greater than their expectations. Located in the center of the Scenic Area, the Museum is located at the point 500 meters away from the North Entrance to the Longmen Grottoes. The Scenic Area would receive almost 2 million tourists on an annual basis and the maximum number of tourists amounts to 7670 per day, including many overseas tourists. However, the emergence of a new building, even if it is only 10,000㎡ and looks small against the huge Longmen Grottoes with a history of 1,500 years, will have great influence on the surrounding environment. Then how should we cope with the reality that the architecture with "4 conformities" has already had influence on Longmen Grottoes? What should we do to make the influence positive and correct? After all, the architecture's influence on the Longmen Grottoes is accomplished by her form, size, materials and temperament. Then how can we judge the attributes of these few elements? These questions lead to the third part of the motive issue, namely, "the architecture design principle".

上: 龙门石窟博物馆区域图;下: 龙门石窟
博物馆景区正门图

Above: Longmen Grottoes regional maps;
Below: The main entrance of Longmen
Grottoes

"设计理念"这个词，从几乎所有设计人从事设计工作之始便一直挂在嘴边，至今以推广普及至每一个与设计工作沾边的人的口中，当然也包括业主与政府官员了。但是，什么是"设计理念"呢？有没有一个明确的词解呢？这个词在不同的人心目中都有自己的解释，可以说是"因人而异"。而有趣的是，它面对不同的对象时也会受对象的影响，产生变化，可以说是"本体"与"客体"之间的一种关系。"设计理念"的书面解释是——"设计师在作品构思过程中所确立的主导思想，它赋予作品文化内涵和风格特点。它不仅是设计的精髓所在，而且能令作品具有个性化、专业化和与众不同的效果。"而我们通过"龙门博物馆"的设计过程发现，这一概念是有可拓展的空间的，正如前文所说，它是上升到哲学高度的"本体"与"客体"的一种对话关系，而"设计理念"的标准答案中的所谓"主导思想"只是"本体"即设计师在对话中的主观意志，而"客体"的思想与意志所发生的作用，是不断在与"本体"的对话中浸透进来，进而改变"本体"的。所以说，"设计理念"不是在方案之初就主观形成，并贯彻始终的不变的思想，而是在设计过程中不断完善与进步，不断的自我审视与否定，是主观与客观的辩论的结果。这也就是为什么，很多设计师在自我成长的过程中，对之前设计方案中的理念感到不满意，甚至觉得幼稚的缘因。因为这种幼稚是"对话"不足的结果，是对"客体"的了解不够造成的。人的主观意志是不可靠的，是自身发展与成长的一个瞬间而已。日本著名企业家，缔造了两个世界五百强企业的稻盛和夫说过"所谓人生，归根到底就是'一瞬间'，一瞬间持续的积累，如此而已"。每一个"一瞬间"都是当时的主体心智与思想和政治、社会环境存在紧密的外部关系之间的互动，而这些外部关系是永远在变化的。在佛教当中，这种外部关系与自我心智的存在就是"实有"。去掉实有就是所谓"性空"。到现在，我们以越来越接近佛教中的精要了。

其实设计理念的形成，正如禅宗的"顿悟"。印度佛教的思想体系如果在中国一成不变的话，恐怕难逃其在本土的命运，在中国也早已消失殆尽了。它是随着与中国人文思想的发展与互动，不断上升到新的层面，不执著于外部事物的相状，从而灭除妄念，"顿悟"成佛的。而佛的原义是"觉者"。这不正是最终觉悟的设计人所追求的吗？

而设计人面临的最大敌人其实就是自己，是自己的"无意识"，设计的过程就是与历史、现在及未来之间寻找位置的过程，去除无意识的"实有"，建立"觉悟"的"缘起性空"的新意识。

这样看来，原先的理念观就是执著于现实的限制中的不觉悟了。更何况，大多数的人把理念变为一种定式，一种简单的形象化的背景画面，一种媚俗。这种理念是经不起推敲与发问的。正如米兰·昆德拉在《生命中不能承受之轻》中的话："在极权的媚俗王国，总是先有答案并排除一切新的问题，所以极权的媚俗的真正对手就是爱发问的人，而问题就像裁开了装饰画布的刀，让人们看到隐藏其后的东西"。

现在，我们再看看建筑设计界的现状，刚刚过去的20世纪可以说是建筑的世纪。新到来的21世纪目前看来也是难逃其魔掌了，只是加上了"数字化"的外衣。而相对于其它领域在这时期所取得的成就，建筑领域的现状可以用堕落来形容。建筑师们各自以积极或消极的"自以为是"在尽情地表演，这当中当然包括我们自己。英国著名作家，艺评家约翰·伯格在其自传性小说《我们在此相遇》中说"我们生活中的每一个人其实都像是舞台上的表演者一样无厘头说着单人脱口秀。我们活在这个世界上总是费尽九牛二虎之力去制造欢乐，讨好别人也愉悦自己，似乎只有这样我们才能活下去，但这样活到最后，我们又得到了什么呢？……"

我们表面上取得了巨大的成功，而我们在共同打造一个"极权的媚俗王国"，我们不断的披上不同的华丽的外衣，并挖空心思找到理论根据，把自己的意图正当化，以期望渡过建筑建成之前的品评，并逃避对可能的结果负责，最终回避建筑本身的责任。我们在躲避那把裁开画布的刀，以及那之后的虚弱无力。我们极尽所能的创新行为，是妥协于批评的，回避责任的助纣为虐。建筑师被认为理所应当要摆出一副时刻准备更新已有形式的辩证姿态……而"灵感与创造是一个陈腐形式到下一个陈腐形式之间呼出的一口气——庞德"。

The term "design principle" is often talked of by all the designers and those involved in the design, including the owner and government officials of course. But what on earth is the "design principle"? Is there a well-recognized definition of the "design principle"? Actually, it is interpreted differently by different people. Interestingly, it changes as the design objects change, which signifies the interrelations between the "identity" and "object". A written definition of the "design principle" goes like this, "It refers to the dominant idea established by the designer when conceiving his/her works and giving the cultural meaning and styles to the design. It is not only the essence of the design, but produces the personalized, professional and unique design result." While designing the Longmen Museum, we find the above definition can be expanded. As mentioned above, the "design principle" regulates the interrelations between the "identity" and "object". And in the written definition of the "design principle", the so-called "dominant idea" is just the subjective will of the "identity", or the designer. Nevertheless, the will and idea of the "object" seep in slowly and change the "identity" as it keeps connecting with the "identity". Thus, the "design principle" is not a subjective creation formed at the beginning of the design scheme that remains unchanged. Instead, it keeps improving in the entire process of design, constantly reviewing and overthrowing itself. It is the result of the debate between the "subjective" and "objective". This explains why most designers would find their previous design principles dissatisfying or even naïve because they are generated by insufficient "dialogues" or less understanding of the "object". The subjective will of the people is not reliable and only constitutes a short instant in the growth period. The famous Japanese entrepreneur Mr. Kazuo Inamori, who created two Fortune 500 companies, once said "Life is just a flash of moment and the wealth made in such a flash of moment. That's all." In each "flash of moment", mind and idea of the "identity" interacts with the close external relations of the political and social environment, which keep moving all the time. In Buddhism, the existence of such external relations and the mental ego is called "tangibles". And the things beyond the tangibles are "emptiness". Now we are closer to the essence of Buddhism.

As a matter of fact, the formation of design principle is quite similar to the "epiphany" in Zen. If the Indian Buddhism ideology had remained unchanged in China, it would have been forgotten by the history as it did in its origin country. But it did evolve itself as the Chinese humanistic thoughts developed. To be specific, it was never confined to its external form and became upgraded after the "epiphany". Actually, the original meaning of Buddha is "the enlightened". And isn't it the pursuit of the designers?

And the greatest enemy of the designers is the designers themselves and their "unconsciousness". The design is a process finding a location among the history, the contemporary and the future. And the purpose of design is to eliminate the unconscious "tangibles"and to establish the new consciousness featured by the "enlightened emptiness".

So the original definition of the design principle can be interpreted as the un-enlightened confined by the reality constraints. Even worse, most people would turn the principle into a set, a simply visualized background scene, or a kitsch, which will not stand against any doubt or questioning. As Milan Kundera said in his the *Unbearable Lightness of Being*, "In the realm of totalitarian kitsch, all answers are given in advance and preclude any questions. So the genuine opponents of the totalitarian kitsch are the question-lovers, whose questions reveal us the hidden things just like the knife cutting off the decorative package."

Now let's have a look at the current situation of the architecture community. The 20th century that has just left us was a century of architectures. And it seems that the 21st century will be the same, simply disguised by the "digital architecture". Compared with the achievement made in other arenas, the architecture was actually declining. The architects, including us, are making their performances in a self-righteous way, actively or passively. The famous British writer and critic John Berger said in his biographical novel *Here is Where We Meet*, "Each of us in our life is performing a single-person talk show on the stage. We are making painstaking efforts to create fun for others and for ourselves. It seems that only by doing so can we live on. But what have we got at the end of our life?"

Seemingly, we have made great success. But the fact is that we are working to build "a realm of totalitarian kitsch". We keep putting on the gorgeous King's Costumes and working hard to find the theoretical foundation to justify our intentions, in a hope to pass the evaluation and avoid being held accountable for the possible consequences and for the buildings we design. Indeed, we are staying away from the knife cutting off the decorative package and the following weakness. Our innovative acts actually compromise with the criticism and avoid the responsibility. Just as Pound said, "The inspiration and creation is just a puff between one old form and another".

好了，看来对建筑的责任问题需以后再专题讨论了，那么我们在龙门博物馆的设计当中应当以何种理念为基础进行工作呢？正如前文说述，我们自己也是经历了"无意识"的过程。当机会飞到眼前时，首先映入我们脑海的，是作为年轻的建筑师我们将多么幸运地通过龙门博物馆的设计成功，甚至成名。脑子中多个著名的建筑，特别是博物馆建筑不断地跳出来，引诱着我们，我们马上就屈服了。我只想形容一下我们今天的建筑师或者说设计工作者共同具有的生活态度，这种态度直白地讲叫做"投降"，含蓄地讲叫"软弱"。就这个问题我会专题来另写一篇文章，讲述我们前后重大的思维变化的过程，这里专门写理念的形成吧。最终龙门博物馆的特聘艺术总监，来自郑州的著名设计人肖艳辉老师提醒并引导我们清醒过来。虽然他没有明确地告诉我们该怎么做，但他以其独特的智慧与超脱的心境否定了我们之前的思路，提出设计的理念要来源于对龙门石窟及佛教艺术的挖掘，对中国古典人文精神的体察，这不正是在告诉我们，"本体"与"客体"的对话吗？为什么中国古代艺术创作的概括与洗练，表达出更丰富的艺术内涵呢？虽然当时苦恼了好长一段时间，才模糊地领悟其中的奥秘，远没有到觉悟的地步，但现在想起，这是人生中最重要的一次蜕变吧。反而庆幸原有思路没有被轻易地接受，从而失去从哲学的高度对待工作的最好时机，在与中国传统哲学的要义如此接近时，却又擦肩而过，那将是人生多么大的损失啊。另外，肖老师的提醒还为我们之后的行动力与结果之间无意地注入了新的动力，那就是激发了我们的"灵性"。关于"灵性"，当然也有程度的问题，我无意言明我们多么具有灵性，然而没有这次与龙门的邂逅，我们与之恐怕将很长一段时间隔着一层窗户纸吧，也许永远也捅不破的，就是这些许的障碍。看似毫不费力的位移，却如同时空的距离，要以光年来计了。

Well, it seems that we have to discuss the issue of responsibility in another paper. Then on what principle should we base the design of Longmen Museum? As mentioned above, we also experienced the moment of "unconsciousness". When we were given the opportunity to design the Museum, the first thing coming into our mind was that, as young architects, we would make success and become famous in the task of designing the Museum. Also coming into our mind were a few successful architectures, especially those of museums, which kept alluring to us till we yielded to them. I am just trying to explain the generic attitude shared by most of the designers or architects in the modern times. Such attitude, if placed explicitly, can be interpreted as "surrender", or "weakness"in an implicit way. About this issue I would write another paper explaining the major changes in our thought. But here I would focus on how the design is formed. Indeed, it was Mr. Xiao Yanhui, the Chief Art Director of the Longmen Museum, a famous designer from Zhengzhou, who instructed and awakened us. Though he has never told us explicitly how we should do, he did deny our previous ideas using his unique wisdom and detached state of mind. He insisted that the design principle be originated from exploring the art nature of the Longmen Grottoes and Buddhism, and the humanistic spirit of the ancient Chinese people. By doing so, he was reminding us to highlight the talk between the "identity" and "object". Why could the ancient Chinese artists, with their concision and concentration, produce the intensive and extensive connotations in their works? Indeed, we were greatly puzzled by this question at the beginning before learning something ambiguously. Honestly, we were not completely enlightened at all. Looking back at what we were at that time, we find it was really an important transformation in our life. Fortunately, the original thought of us was not accepted in an easy way. Otherwise, we would have lost the opportunity to do our job in the perspective of philosophy. If we had missed the essence of the traditional Chinese philosophy when we were so close to it, it would have been the greatest loss in our life. Furthermore, Xiao's instructions, unconsciously, instilled the new vitality into the subsequent power of action and results, which triggered our "smartness". I am not saying how smart we are. But it is the truth that without taking the job of designing the Longmen Museum, we would never meet our smartness, though it was only a few meters away from us. And these few meters, if measured in the geographic context, is actually a few light-years.

二. 设计理念阐述
II. Elaboration of the Design Principle

龙门博物馆的设计理念分五个部分，分别是

    （一）龙门——无可替代性的空间。

    （二）可视世界与理想世界的过渡空间。

    （三）龙门石窟是反建筑的空间。

    （四）佛教哲学与艺术的"空、圆"。

    （五）"心性和直觉"的情境空间。

（一）龙门——无可替代性的空间

    第一部分是就博物馆的功能及所处位置与龙门石窟的整体外部环境的关系进行分析与提炼的。如前文所述，理念是"本体"与"客体"对话过程的结果。"本体"——设计师的主观意志是个变量，有诸如成熟、深刻及幼稚、肤浅之分。而"客体"——龙门石窟及佛教艺术是现实是不变量，但也是无限量，浩如烟海，过于庞杂。设计理念在第一部分就是要在这浩瀚的佛国中找到线索，龙门的线索，佛教艺术的线索，和建筑师的意志与思想碰撞的线索。其实这也是龙门自身性质与气质的提炼。就此而言，便引出"性"与"相"的关系问题。"性"是性质，"相"是表相，"透过现象看本质"就是本条理念的目的。

The design principle of Longmen Museum contains the following 5 parts:

(I) Longmen – the irreplaceable space;

(II) The transition from the visible world to ideal world;

(III) Longmen Grottoes is the anti-architecture space;

(IV) The Buddhism philosophy and the "emptiness and interpenetration" of art;

(V) The context space of "disposition and intuition".

(I) Longmen- the irreplaceable space

The first part analyzes the interrelations between the functions and position of the Museum and the overall external environment of the Longmen Grottoes. As mentioned above, the design principle is the result of the talk between the "identity" and "object". Here the "identity", or the subjective will of the designers, is a variant. Sometimes it is profound and mature and sometimes it is shallow and naïve. However, the "object", or the Longmen Grottoes and the Buddhism art are the invariant and unlimited, which are much too complex. The first part of the design principle tries to locate in the vast realm of Buddhism the clues to Longmen and the Buddhism that overlap with the will and thought of the architects. Indeed, it is extracting from the nature and temperament of Longmen. Based on this, we have reached the interrelations between the "nature" and "look". "Nature" refers to the property while "look" means appearance. So this part of the principle can be interpreted as "seeing through appearance to perceive the essence".

"相依性而现，性由相而彰，性相二者，一表一里，从不分离"。佛教哲学已经精辟地指明了性与相的关系，而"相"是可以复制的，"性"是不可复制的。

龙门石窟从"相"上来看，其独特的空间语言是大家都极为熟悉的。目前的龙门是2000多个洞窟，大大小小，深深浅浅的散布在石壁上，其间看似随意，其实精心组织与经营过的石阶，将大部分洞窟连在一起，供游人及香客通行与瞻仰。如果我们利用这极具特色的空间肌理，有选择地在博物馆上使用，一定会让人联想到石窟的大背景，这显然是一条设计之路，然而这样是正确的路吗？或者说是最佳的路吗？现代主义的设计作品中不乏案例，以绚丽的空间组织，与交错的构成线索组织建筑，然而这些肌理是佛教艺术的本质吗？

首先龙门石窟从北魏开凿至宋代，经历了801年的历史，是由一至多，由点至线，逐步完成的。不知有多少人，多少财力物力投入到这个宏大的工程当中，每一个洞窟都是血汗的见证，每一条石阶都是岁月的页码。时至今日，又历经无数年的风霜雪雨的洗礼，再给龙门石窟罩上了一层时间的代码，而时间是人与历史与未来的空间。我们在龙门建造的是1500年来最大的建筑工程项目了。而且我们可能只需半年就要完工了。虽然前后加上设计与克服种种困难去说服各级领导，以及开工建设已6年有余，但是与龙门相比，这也只是一个瞬间而已。

建造一个建筑物的时间是极其短暂的，短到我们可以称之为瞬间，与此相比，建筑物一旦建成后，它持续使用的时间可就是漫长了，这是建筑物的一个重要特性。在它使用的过程中，会与龙门一同存在，与龙门对话。而它有这个资格与理由和伟大的龙门石窟一同存在吗？它会说："嘿！大哥，我是你的微缩版！"。而龙门只是"拈花微笑"而已。因为在她眼中，这个小怪物连一个跪伏在地的香客都不如，充其量是个穿着"到此一游"T恤的游客，还讨厌的长久地占据了她的视野。如果龙门博物馆是她1500年来等候的朋友，那这个朋友将会多么的荣幸啊，我们打造的建筑物应当配得上

这个身份，应该是个与龙门能促膝长谈的朋友，从新朋友到老朋友，互为知己的朋友。龙门博物馆不是龙门石窟艺术珍品的新仓库，而是介绍和推荐老朋友的人。她首先应当了解龙门石窟，了解她的"相"，更了解她的"性"。而她也应有自己的"相"，自己的"性"。这才是我们的博物馆。

作为一个老朋友，我们应当首先思考——"龙门石窟"是怎样产生的。这当然不止是说是由谁发起并开凿的，而是指更重要的，是什么力量支持着她几百年来不断的扩展，1500年来历久弥新的存在与灿烂呢？下面的答案带出我们本理念的第一要点……

"The look is presented while basing on the nature, which is shown through the look. The nature, lying below the look, never separates from the look". The Buddhism, in simple words, has told us the interrelations between the nature and the look. The look can be duplicated while the nature can't.

Many people are quite familiar with the "look" of Longmen Grottoes because of her unique space features. At present, it is composed of over 2,000 grottoes, big or small, shallow or deep, which are scattered on the cliff. The seemingly random stone steps connect most grottoes in a well-organized way for the tourists and Buddha pilgrims to reach them. If we use the typical space texture selectively in the Museum, it would remind the visitors of the overall picture of the grottoes. Evidently, this idea is an option. But is it correct, or is it the best? Indeed, many such cases can be found in the modernism architectures, which are structured by the gorgeous space and overlapping clues. But is such texture the essence of Buddhism art?

First of all, the Longmen Grottoes was completed in 801 years from the North Wei Period to Song Dynasty in a slow and gradual way. Many labor forces and wealth had been invested in this splendid project. Each grotto witnessed a large amount of sweat or even blood. And each stone step left a page number in the chronicles. The long history has covered the Longmen Grottoes with a layer of time code. And time constitutes the space between the past and the future. What we were going to build in Longmen is the largest construction project even seen since 1,500 years ago. And this project could be completed in only half a year. Though the design, lobbying efforts and construction work took us 6 years, it was just a flash of moment compared with the 1,500 years' history of Longmen Grottoes.

Though building an architecture only takes us a flash of moment, the architecture has to last long. And this is an important feature of all buildings. In the lifecycle of the Museum, it will coexist with the Longmen Grottoes and keep talking with her. But is the Museum qualified enough to stand side by side with the Longmen Grottoes? It will say to the Grottoes, "Hey, big brother, I am a miniature of yours". But the Longmen Grottoes would simply show a Buddha smile. In her eyes, this weird little thing is no good than a kneeling Buddha pilgrim. At most, it is a tourist wearing a T-shirt with the "I am here" slogan on it. But how dare it stay in her eyes for a long time? If the Longmen Grottoes regards the Museum as her friend, how honored this friend would feel? And we should design such a building as a friend qualified to talk with the Longmen Grottoes and growing from a new friend to an old one. The Longmen Museum is not just a new warehouse storing and displaying the art treasures of Longmen Grottoes. Instead, it should be a friend of hers introducing new and old friends to her. To do this, the Museum should first understand the Longmen Grottoes, her "look" and her "nature". On the other hand, the Museum should have its own "nature" and "look".

As an old friend, we should think how the Longmen Grottoes has become what she is like today. We do not mean how the project was launched and who carved her. What we want to know is what strength has supported her constant expansion, fresh existence and eternal brilliance over the past 1,500 years. The following answer to this question will lead to the first key point of our design principle.

信仰的力量
The strength of belief

是信仰的力量造就了龙门石窟，以及她灿烂的佛教文化。从浩大浩大的工程量，从卢舍那大佛伟大的圆融，从万佛洞精微的雕琢与无畏隐忍的耐心，从那饱浸着血泪的石阶，所有的细节都在诉说着这样一个本质——是对佛教信仰的无尽的虔诚与热爱，带给历代帝王与信徒力量，满怀热情、充满期望的完成了这一伟大的工程。没有佛教这一信仰，就没有龙门的开凿，而没有了这份虔诚，就没有龙门的伟大。如果说龙门石窟是工匠们一斧一凿地雕琢出来的，不如说是信仰的血液涌动下，虔诚地生长出来的，而且还将继续生长下去。而我们生活的当下，是个多么缺乏信仰的年代啊，人们好像早已失去了虔诚与宽容的能力，只是满足于眼前的物质利益，成为欲望的奴隶。就建筑业来说，包括我们自己在内，建筑往往不再是一种理想，而是混饭吃的工具了。诱惑我们的东西太多，"新、奇、怪"的轰动效应；做出别人之所不能，而不考虑别人是否只是不屑；穿上时尚的外衣，并神经过敏的总觉得还是掉了队；把高科技当成救命草，电脑成为兴奋剂；数码时代中，人们的眼中只剩下像素……

龙门震撼了我们的内心，与龙门谈心的过程也完成了我们的自我救赎，就像"佛"的原意并非是什么神灵，而是"觉者"。哲学性也是任何一部伟大的宗教的共性。这也是宗教之所以能够长久存在的理由。我们不必为了虔诚就把自己变成佛教的信徒，我们应当把自己融入到她的哲学与文化中去，因为佛教的哲学是去除"实有"，去除"杂念"，不执著什么，才能"觉悟"。所以对于普通人或无神论者来讲，人生的信仰应当是对人生哲学对智慧与觉悟的追求，把觉悟变成理想吧。

当我们把成功与成名错误的当成理想时，我们就已经走在扑朔迷离的道路上了。佛教的信徒一代一代费尽一生的心力去完成龙门的雕琢，虽然自知这也只是这一伟大工程的一个细节。但他们知道他们的后人仍将继续，他们要做的只是"赋出"。"赋出"是一种美德，"专注"是一种修炼，他们雕琢的不是洞窟，他们雕琢的是他们的理想，因为"只有理想能波及四海"。—— 密斯·凡德罗。

现代，我们拥出高新技术，工程机械，我们可以修建无比复杂与宏大的建筑物，而我们能够再实现一个龙门石窟吗？我们什么都有了，却偏偏少了理想。

反观如今的龙门，可能有人会问，她昔日的辉煌是否已失，无数工匠的创造已不复存在，那他们的信仰与理想在哪里，这就是本理念的第二个重点……

空间是精神的载体

的确，历经1500年走到今天的龙门，大部分精美的雕像已遗失与惨遭破坏，大部分洞窟空空如野，在里面的似乎只有空气。曾经精美华丽的石窟寺，是布满壮丽的木构斗拱与飞檐的，如今也与战火一起灰飞烟灭了。不变的是，今天的龙门石窟依然充满熙攘的人流，只是手里拿的不再是香烛，而是相机罢了。

回想起当年站在珠峰脚下时，望着无际苍凉的群山，突然感到自己是多么的渺小与无助。因为自己被空无所征服。然而睁开眼又

The Longmen Grottoes was actually made with the strength of belief and the brilliant Buddhism culture. The huge amount of workload, the great interpenetration of the Grand Locana Buddha, the meticulous carving and fearless patience of the 10,000-Buddha Cave, the stone carvings immersed with tears and blood, all these show the great piety and passion for the Buddhism belief, which brought strength to the Emperors and common believers who finalized the great project with their bravery and expectations. Without the Buddha belief, there would be no such thing as the Longmen Grottoes. And without the piety of the carvers, the Longmen Grottoes, even after it was built, would not have been so great. The grottoes were not carved by the craftsmen holding their chisels and other tools. Instead, they were planted and fertilized by the piety. The era we live in is in short of belief and our people have lost the capacity to show piety and generosity. They are simply satisfied by the immediate interest, thus becoming the slaves of all kinds of lust. In the community of architecture, including us of course, architecture is no long an ideal. Instead, it is just a tool to make a living. Too many allures are here in front of us, including the sensational effect produced by the "original things", doing things that other architects wouldn't do, putting on the most fashionable clothing and still worrying about being out of fashion, only resorting to the advanced technologies, in particular the computers, or having only pixels in the mind in the digital era.

Longmen gives a great shock on our mind. And the talk with Longmen has helped us to realize the self-redemption. Because the "Buddha" in itself originated from an "enlightened person" instead of a god. Indeed, all religions have the nature of philosophy and that's why the religions often exist for a long time. We don't have to turn ourselves into the Buddhist believers to show our piety. Instead, we should indulge us into the philosophy and culture of Buddhism simply the Buddhism philosophy advocates eliminating the "physical things" and "distraction thought". Only by getting out of the clinging can we get enlightened. So for the common people and atheists like us, we should pursue the philosophy of life, wisdom and enlightenment, and work hard to turn the enlightenment into our ideals.

So when we are mistaking success and game as the ideals, we are stepping on the way to obsession. The believers of Buddhism worked in Longmen on the grottoes for generations, though they fully understood what they had done was just a tiny part of the entire mission. They knew their future generations would continue their work and they just kept contributing their days and nights, regarding it as a virtue and their focus as a practice. They were not carving just a grotto. Instead, they were carving their ideal because Mies van der Rohe once said, "Only ideals can be spread all across the world".

In the modern times when we already have the advanced technologies and the engineering machinery, we can build the most complex and splendid buildings. But can we build another Longmen Grottoes? We have had everything in our pocket, but not the ideal.

Someone might ask if the past splendor of the Longmen Grottoes has all been lost. Since the creation of the numerous craftsmen is no longer staying with us, where are their belief and ideal? And this is just the second point of our design principle?

The space is the carrier of spirit

As a matter of fact, the Longmen Grottoes, which has reached us from 1,500 years ago, had most of its exquisite sculptures lost or damaged, leaving nothing in the grottoes but air. The Temple that used to be gorgeous and magnificent with splendid decorations on the roof had all disappeared in the past wars. Nonetheless, one thing remains unchanged. The Longmen Grottoes is still crowded with groups of people, who are holding cameras but not the incense.

One day when I stood at the seemingly boundless and desolate Mount Everest, I suddenly felt how small and helpless as if I had been conquered by the emptiness. But my eyes were then attracted by an old woman and a small kid kneeling-walking on the path. Compared with the boundless mountains, they were so small in size. However, the shock they produced on me filled in almost all the views in my eyes. They were showing the fearless pursuit for their belief. And I, taking a camera and standing beside them, looked totally incompatible with and redundant in the scene and atmosphere. Looking down on my feet ashamedly, I found I was just standing at the same place as the other tourists would to have a distant look and take a picture. Unsurprisingly, lying on my feet were the coke tins and packages. The garbage produced and left by the human civilization had invaded into such a distant place. And it seems the Longmen Grottoes is suffering from the same invasion. Different from Tibet, the Longmen is fighting against more aggressive invaders. And she has to conceal the "nature" in the empty caves by utilizing the brilliance of Buddhism, and presented her "look" to the cameras, saying to them, "Hey, keep an eye on your bag". Probably the eternal smile of Grand Locana Buddha is including some contempt in it.

一个匍匐跪行的老妇及步行跟随的只几岁大的孩子所吸引。与空旷的群山相比较，这两个人小到可以忽略不计，但是他们带给我的心灵震撼又充满了我目之所及的空间内。追求信仰的精神，是多么的无所畏惧。倒是站在附近，端着相机的我，显得那么的不谐调以及多余了。当我惭愧地低下头，却发现自己站的位置是猎奇者经常停车眺望与摄影的地点，自己一点儿也没有脱俗地站在几个可乐瓶与包装纸的尸体上。如此人迹罕至的地方，也已经被文明的垃圾所侵略了。这多像现在的龙门石窟啊，只是龙门对抗的侵略要跋扈得多而已。也只能把佛教的光辉，把"性"藏在空荡的洞窟里了，并且无私地把"相"奉献给相机们，并对相机们说"嘿！别光顾照，看好包！"卢舍那永恒的微笑此时也是略带些轻蔑吧。

　　刚去龙门时，我们自己也是相机们的一份子而已。但是在我们为存储空间不足而苦恼时，还是无奈地发现丝毫没有捕捉到龙门的精神内涵，干脆丢掉相机吧。其实真正的内心变化才刚刚开始，因为沟通渠道的改变，才是走向正确方向的基础。现在写文字时，才最终领会龙门需要的不是精彩瞬间，因为这些所谓的精彩对她而言无非是飞扬的尘埃，除了迷眼外似乎一无是处，她早已尘埃落定。虽然稳定了1500年，可还有下一个1500年呢，甚至更长。在外部世界瞬息万变之时，浮躁的情绪无所不在，这时候静下心来冷眼观察的人才是赢家。

　　我们开始冥想，或者说发呆。走到伊河对岸，对着龙门的全景，从中远距离再看龙门时，细节变得不再那么重要了。

　　宏大的全景让人回复到了整体的空间层面。前文说过，龙门像是生长出来的，在一种能量的驱策下。这种能量就是信仰，佛教的信仰。而人们为什么选择用最困难方式来构筑理想呢？在平地上建造不是更容易？答案是真诚与永恒。人们在山壁上雕琢的信仰，需要以更真诚更虔诚的心来支撑，困难这时不是障碍了，相反，是挑战，是克服困难后的心灵升华，是真正的快乐。相信在珠峰脚下看到的老人与孩子，当他们活着到达拉萨时，这一路的风餐露宿都变成幸福的笑容。因为他们是以生命为赌注来完成这一使命的。也许会有人笑他们太认真，太愚笨，可我看到的是惭愧。佛教在中国的发展一部分是种进步，另一部分是妥协，还有一部分是放弃。禅宗发展的结果是"萧然静坐，不出文记，口说玄理，默授与人"。最后是连字都懒得写，话都懒得说了，更别说再去西天取经了。当然这里不是在嘲笑禅宗懒，禅宗是接受了"道"与"儒"的思想后，产生思想变化的。其行为特征是"不执著"，对形式的无限放宽的结果。但普通中国人的思想中，充满着"空、无、虚、幻、道"等虚无飘渺的东西，甚至认为才是有思想。对待执著的行为又有例如"儒、释、道"的"去除执著与实有""顺其自然""虚无""无为""中庸""和"等等理论基础，所以，中国人就爱笑别人认真。

　　龙门石窟就是认真出来的结果，佛教艺术中的精品也是认真的结果。这种认真本身就是一种精神，它不以洞窟的保存现状为衡量标准。佛像固然没了，精神是永存的，反倒是这种"无"的现状，引起人对其"有"时的联想吧。特别是卢舍那大佛周围的宏伟雕像群，屋宇烧掉了，肢体残缺了，而空间还在，反而因局限其空间的构筑物消失，释放出更大的空间。无论是尺度上的，还是精神层面的，这就

When we just came to Longmen, we were simply parts of the cameras. But after finding we were running out of our memory disks, we were still not able to capture the spiritual connotations of the Longmen Grottoes. Then why not forgetting about the cameras? At that point, we began to experience the change in our inner mind and started to be steered to the right direction because of the change in the communication channel. Now when I am writing this paper, I can say I have understood that Longmen doesn't need the so called wonderful moments because these moments, for the Longmen Grottoes, are nothing but the flying dust. She is now completely sublimated before the dust has settled. She has been there for 1,500 years is waiting for another 1,500 years or even longer years. When the external world is changing busily and noisily, only those who remain calm and peaceful in mind can become the genuine winners.

I sunk myself into the meditation, or trance. Walking onto the bank of Yihe River watching the panorama of the Longmen Grottoes, the details are not significant at all.

The grand panorama takes us back to the level of the comprehensive space. As mentioned above, the Longmen carvings were generated by certain strength. And such strength is the power of Buddhism belief. Why would people select the most difficult way to construct their ideal? Is it much easier to make the carvings on the flat land? The answer lies in the people's sincerity and the eternality of their belief. The belief demonstrated on their carving work on the cliffs needed to be supported by even more sincere and pious hearts. At that point, no difficulty constituted the obstacle. Instead, it was the challenge, the spiritual sublimation and the genuine pleasure derived from overcoming all difficulties. I believe that the old woman and the kid I saw at the foot of Mount Everest, when they arrived at Lhasa after all difficulties including hunger and bad weathers, would have the happiest smile on their faces because they have accomplished their mission at the risk of their lives. Some people might mock at their over-passion or stupidity. But I did feel small in front of them. The development of Buddhism in China is the result of progress, compromise and abandonment. Zen, the typical Chinese Buddhism, advocated passing on the doctrines orally instead of in written. We should never laugh at their laziness. Indeed, Zen's thought has changed gradually after it accepted something from Taoism and Confucianism. It is featured by "not obsession or insistence" and extremely lifting all bans. In the mind of an average Chinese, there are many things that are "empty, nil, intangible, fantastic and Taoist". They believe these things are the real thoughts. In the Chinese philosophy, in particular in "Confucianism, Buddhism and Taoism", there are many theories such as "eliminating obsession and tangible things", "going as it is", "emptiness", "doing nothing", "mediocre" and "harmony". That explains why the Chinese people often don't like over-passion or persistence.

But the Longmen Grottoes is just the result of over-passion and persistence. And the classic Buddhist art works are also the

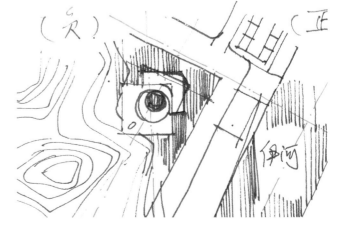

results of such nature. Indeed, the over-passion and persistence themselves are a kind of spirit, which is not measured by the status quo of the grottoes. Even though the Buddha carvings are gone, the spirit is still there, reminding us of the state when the carvings are standing in the grottoes. In particular, though the buildings holding the magnificent carving cluster around the Grand Locana Buddha were built and most Buddha carvings lost their legs or arms, the space is left physically and spiritually. We owe thanks to those making the damage such as in the wars for their mercy of leaving us something. And the builders of the Longmen Grottoes chose the special way and place to build her just to keep it eternal. From this perspective, they are the sole winners. Unless the Grottoes are bombarded to ground zero, the Buddhist spirit will be passed on for generations as eternal as the Sun and Moon.

Now it is time to summarize the first time of the design principle. The Longmen Grottoes was built thanks to the strength of belief. Because the space is the carrier of spirit, it determines the eternality of the Grottoes. Both the spirit and the eternality are about the pious and persistent pursuit of the Buddhist believers for the religion. So what people see in Longmen is the materialized piety.

Let's go back to our Museum. If we just want to build a functional building, it is no use to write this paper at all. But we wish to position us in the same level of the builders, carvers and craftsmen of the Longmen Grottoes, trying to have the similar mentality that they had. Building a project close to the great and ancient site, we fully understand it would be the most important engineering work in this place ever since Longmen Grottoes was built 1,500 years ago.

All buildings will end in two fates. In the first fate, the building is not suitable in a certain years and gets demolished. And in the second fate, it will become part of the people's memory and gets preserved. We won't hope that the Museum will be as eternal as the Longmen Grottes. But we do hope it will become part of Longmen, a segment in the history and play a positive role. If it is not able to reach the above objectives, there is no need to build it at all.

As a building, it has to have some kind of shape, which will definitely lead to the interrelations between the form of existence and freedom, which are actually the opposing elements that dominated the architecture in the latter part of the 20th century. To avoid the rigid form, the architects tended to pursue freedom, which in turn has to be implemented in the way of form. So some balance is needed between the form and freedom. And such balance is called the "liberated form" or the "limited freedom" that have their separate theories. For instance, now the concept of "digital buildings" is accepted by many architects. There are two types of people who keep sitting in front of the computers. Type 1 is the active inventors, composers and artists. Type 2 is the captives of software. But both Type 1 and Type 2 create things, even if they are dynamic

如来坐像
Seated Tathagata Statue

候我们反而要感谢诸如战乱类的破坏者们，破坏时也不是那么认真吧。建造者偏偏选择最难的方式来建造，也是希望她能永恒的存在，难以泯灭。从这一角度说，他们是胜利者，除非把龙门山炸平，否则这团佛教的精神永远会存在下去，与天地同寿，与日月同辉了。

写到这里，理念的第一部分该总结一下了，正是因为信仰的力量造就了龙门石窟，因为空间是精神的载体，决定了龙门石窟的永恒，这两方面都是围绕着对佛教信仰的虔诚与执著的追求。人们在龙门看到的，都是这种虔诚的物化。

再回到我们的博物馆吧，如果我们只去建造一个功能建筑，以上都是废话，何必呢？好好看看就行了。而我们希望和龙门的开凿者那样，以接近他们的心态为标准，珍惜这次机会，在这古老伟大的地方建造。这个建筑将是龙门石窟1500年尘埃落定后的最重要的工程。

建筑只有两个结果，一是若干年以后不再适用，被炸掉；二是它成为人们记忆的一部分，舍不得失去，反而将其保护起来。我们当然不奢望这个建筑与龙门石窟一样永恒，但是希望能成为龙门的一部分，作为龙门历史的片段，尽量久的发挥作用，当然是积极的作用。如果反之，最好现在就不要建。

建造就一定是造型问题，有造型就存在形式与自由的关系。形式与自由，这两种东西互为矛盾，这对矛盾曾经支配了20世纪后期的建筑业，也将在21世纪继续发挥作用。形式为了避免僵化，总在寻求自由，而自由想要成为现实，又不得不落实到某种形式，然而双方总是在一种爱恨中找到平衡。各自取名"解放的形式"与"有限的自由"。然后再寻找理论基础，寻找各自的"主义"，例如目前最流行的"数字建筑"。在计算机前的人有两种，一种是积极的创作者，一种是软件的俘虏，也是机会主义者。但无论哪种人，那种由他们在电脑中创造的充满动感与可塑性的物体，在被固定在现实中的过程都是痛苦的，或者是沉重而无趣的。此时，数字建筑彻底地背离了它的初衷，形态上的"自由"却更加明显地反衬出这类建筑本质上的不自由。其实什么事情都是一个"度"的把握的问题。建筑师一生只是在做一件事——"权衡"。

龙门博物馆的设计过程也是一个形式与自由的权衡过程，但这里自由体现在非宗教的，无沉重历史背景与文化压力的感性思维，形式则是一种斗争了。斗争的对象是佛教艺术主题及龙门石窟的象征性与形式主义的矛盾，国际化与民族性的矛盾，还有就是建筑设计师与评判者的矛盾，最后还有一个，也是最关键的，建筑设计师渴望成功的愿望与对建筑及龙门石窟的责任之间的矛盾。这一系列的矛盾每一项都可以令建筑产生质的变化。幸运的是，作为建筑师的我们，在前辈的提醒与帮助下，在经历了波折后，很快找到了自身的定位，用理念为武器战胜了形式主义的诱惑，与我们自己。

我们首先从信仰的力量中获取了能量，从空间是精神的载体中获得鼓励与解放。龙门石窟的"形"是其"质"的表相，是"相"与"性"的一表一里，相得益彰的表现。博物馆是其质与性的展示窗口，不是其形与相的克隆，因为在一个一万平米的建筑上，无法承载 有1500年历史积淀的龙门石窟的厚重。我们创造的是龙门石窟1500年来的新伙伴，一个年轻的，传承其精神的新个体。具有独立的"性"与"相"。

龙门的辉煌不应被肤浅的模仿，就像世界公园的微缩景观，只是供不愿行万里路的相机们去调戏一样。我们所要做的第一件事，就是用我们的感悟，去"觉悟"所有有可能对我们的工作指指点点的人，龙门的符号性是多么危险的敌人，龙门本身就是为唤醒人们觉悟而存在的，而任何试图对龙门空间的模仿都是在"执迷"地走向万劫不复。龙门的空间是时间与你我之间的空间，这个空间是无法替代的。

and flexible in the computers, which are painful and boring in the reality. At this point, the digital buildings have been distorted from its original essence. The seeming "freedom", in a more evident way, reflects the "not freedom" in such buildings. As a matter of fact, we have to control the level or degree of things. The architects, in their whole life, are doing just one thing, to get access to "balance".

While designing the Longmen Museum, we were also working hard to reach the balance. But in our design, the freedom has nothing to do with the religion. And there was no such thing as the heavy historical background and cultural pressure. Nevertheless, the pursuit of form was quite similar to a fight. And we fought against the conflict between the Buddhist art theme and the symbolism and formalism hidden in the Longmen Grottoes, the conflict between the international feature and the national characteristics, the conflict between the architects and the critics, and most importantly, the conflict between the architects' aspiration for success and their responsibility for the Longmen Grottoes. Each pair of these conflicts, if not well handled, will lead to the fundamental change in the Museum. Fortunately, we were instructed by the experienced architects. With their help and advice, we quickly found the correct way, conquered ourselves and the allures of formalism.

First, we obtained our strength from the belief. Then we got encouraged and liberated from the space, which is the carrier of spirit. The "form" and "look" of the Longmen Grottoes are the external presentations of her "nature". As the window showing the nature of Longmen Grottoes instead of simply copying her look or form, the Museum, which is only 10,000 square meters in size, is not able to take the heavy cultural, historical and religious contents of the 1,500-year-old Longmen Grottoes. What we were built was just a new friend of Longmen and a young individual that can carry on her spirit. The Longmen Museum has its own "nature" and "look".

The splendor of Longmen should never be copied in a shallow way, just like the miniatures landscapes in parks are played by the cameras wouldn't travelling from a long distance. The first thing we had to do is to "perceive" all those who might point their fingers at our work and make unfriendly comments. The Longmen Grottoes, has a dangerous enemy as a sign of religion, because her existence is to enlighten people. And any copying endeavor will come to a dead end of "obsession". The space of Longmen belongs to all of us, and this space is irreplaceable.

龙门石窟门前局部照片

*The partial view of the entrance of
Longmen Grottoes*

（二）可视世界与理想世界的过渡空间

(II) The transition from the visible world to the ideal world

证明了"符号"的不可复制性，空间的无可替代性后，我们又应该用何种方式来表达建筑呢，用形式的表现力吗？用形式背后的思想吗？形势与思想需要或可以割裂开看待吗？这就又回到了形式与自由的矛盾问题上来了。在我看来，形式本身就是思想，对形式的选择本身就是对个性与共性，或"普遍性"与"特殊性"的回答。形式就像一件乐器，除了自身的形态外，还会发出美妙的声响。这声响当然也与其形态有密切的关系，与声学的混响相关。对形式的突破与革新永远局限于那种乐器特有的声音，如果连声音都要突破，那就干脆发明一种新的乐器好了。然而如今的建筑师们却致力于一件件奇怪可笑的乐器形式，其发出的声音却最多是在追赶传统与经典乐器的音质，甚至还要差很多。对于形式与自由的关系，就像乐器对形态自由的追求与无奈一样吧。所以问题不在形态，而在声音。

龙门博物馆要发出什么声音呢？还是要回到龙门石窟，回到佛教中来寻找答案。

We have proved that the "sign" cannot be copied and the space is irreplaceable. But how can we express the architecture? Should we use the expression power of the form, or the thought behind the form? Do we need to, or can we separate the form and the thought? It is again the issue about the contradiction between the form and freedom. In my opinion, the form in itself is the thought. Our choice of the form represents our answer to the generic or specific feature. The form is just like a piece of musical instrument, which has its own form and produces the fascinating sound. And the sound it produces is closely related to the form of the instrument and its acoustic features. No matter how much we want to make new breakthroughs in the form of the musical instrument, we are limited to the special sound it produces. If we wish to make innovations in the sound, we are making a new musical instrument. In this case, the many pursuits made by the modern architects are weird and ridiculous. What they are doing is just like making some innovations on the musical instrument. And the result of their work is just to produce some traditional or classic sound, sometimes even worse than the instrument can produce before it is reformed. The interrelations between the form and freedom are like the same. The key point is not the form, it is the sound.

So what kind of sound would the Longmen Grottoes give out? Let's come to the Buddhism religion for the answer.

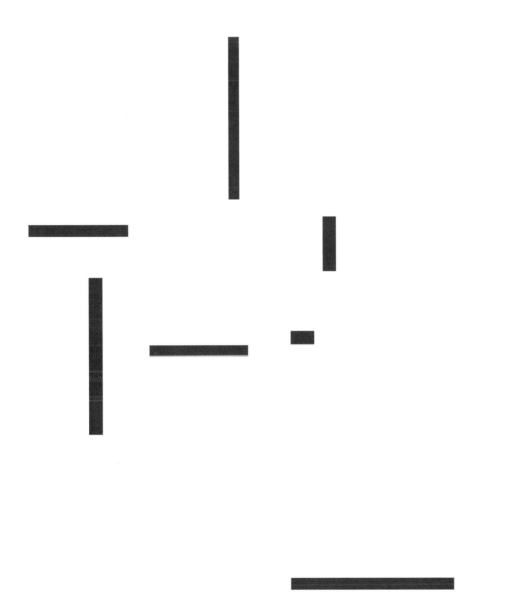

前文提到，当我们站在龙门石窟东山，对望石窟主体的全貌时，感受到了精神的空间。而这空间不只是在全景中缩小为各种尺度的洞穴与线性的石阶连接构成"点、线、面"的平面空间，也不只是随山壁起伏排列并深入到洞穴深处的三度空间，而是一个历史、现实与理想之间的四度空间。

当你面对龙门时，如果假想一下你不是在现实世界，而是在一个佛教的精神世界中，周围不再是游客与相机，也没有数字化与模拟化的纠结，只有佛教的"性空"世界，你所要做的只是探寻心性的本源。你的周围只有无尽的圆融。而这个世界就在龙门石窟大大小小的洞穴的背后。洞穴就变成了现实与佛教理想世界的沟通媒介，变成了佛教世界的窗口，多么有趣的一个比喻。在内部的我相对外部世界一切太喧嚣的事业和太张扬的感情都产生了怀疑。它们总是使我想起莎士比亚对生命的嘲讽"充满了声音和狂热，里面空无一物"。

想到这里，我们的博物馆就有了一个可以类比的定位了。我们建造的不是现实世界的嘈杂与喧闹，也不是清明宁静的佛国乐土，而是由现世步入佛教理想世界的过渡空间，这里面有几个不同的层面。

一是将佛教世界以外的人引入佛教的世界，了解佛教，认识佛教。

二是将信仰佛教的人引入更高层次的佛教哲学与艺术世界，

超脱出求神拜佛的低级阶段。

三是引导参观龙门石窟的观众，在实际进入龙门石窟之前，了解龙门，认识龙门，成为一个功能的过渡空间。这个空间是博物馆的真正的功能定位了。虽然"博物馆"顾名思义是展示众多"物"的馆，然而博物馆真正的意义在于教化民众，提高认知，所以博物馆的功能是进与出的思想过渡。我们真正关注的不再是展品的数量与展厅及流线的分布，这些最多是第二位。真正应该去做的是展品的总合与空间本身对参观者的影响力。这样，我们从关注龙门石窟的自表及里，到了建筑空间重在内部影响力的思维过渡。而这种影响力就是前文提起过的"声音"了，决定了发出的声响，再来说乐器的形式吧。

As mentioned above, when standing some distance away from the Longmen Grottoes and observing her panorama, we felt the space of spirit. And this space doesn't mean the space composed of the dots, lines and surfaces, or the caves connected by the stone steps, it is the history and the four-dimensional space between the reality and ideal.

Standing in front of the Longmen Grottoes and imagining you are in the spiritual world of Buddha instead of the reality world surrounded by tourists, cameras and the annoying dilemma between the digital and analog data, you will focus on pursuing the origin of your mind in the "empty" world of Buddhism filled with interpenetrations. This "empty world" just lies behind the caves in the Longmen Grottoes, which is now a media connecting the reality with the ideal world of Buddha. Standing outside this world, we begin to doubt the meaning of the so-called successful career and the human emotions. It reminds me of a ironic comment made by Shakespeare, "Full of sound and frenzy, it was empty".

Now our Museum has a position. What we would build is nothing about the noise and chaos in the reality world. Nor is it about the peaceful and tranquil Buddha world. It would be a transitional space from the reality world to the ideal world of Buddha. And there are several layers in this positioning. First, the Museum will attract the people out of the Buddha world into it, helping them know the Buddhism. Second, the Museum will lead the people already in the Buddha world to a higher level of Buddhism philosophy and art world, helping them out of the junior level of worshiping the Buddha. Third, the Museum will help the visitors and tourists have some knowledge of Longmen Grottoes before they see it. Thus, the Museum is a transitional space.

This transitional space is the accurate positioning of the Museum. Though the Museum should be a place displaying the items, its greater meaning lies in educating the people and raising their awareness. So the function of the Museum is to guide the people in and out of their thought. We shifted our focus from the number of exhibits and layout to the impact of the Museum space and arrangement on the visitors. In this way, we have touched the essence of the Museum. And we have come back to the "sound" mentioned above. After deciding what kind of sound we wish to produce, we will think about the form of the musical instrument.

（三）龙门是反建筑的空间
(III)Longmen is the anti-architecture space

上：龙门石窟全景照片；右侧：龙门石窟局部
洞窟
Above: The panoramic view of Longmen Grottoes;
Below: The partial caves of Longmen Grottoes

当我们构建建筑时，无论何种情况，首先思考的是一个体量与外部的关系问题。例如，是如何围合出"场"。墙的意义是界限与私密，是场所感。当然，场所感不一定只靠墙，任何能对空间产生影响的东西都可以。但墙是最简单直接的方法了。

然后是柱，然后是出入关系，再然后可能是层与层之间的空间与交通的联系了。形式与表现力因人而异，不必多说。所以大部分或全部传统意义上供人使用的建筑都大抵如此。窗与墙是分割内外空间的手段。外面是空，里面也是空，区别是为何而空而已。再来看看龙门石窟，如果我们把刚才直面龙门西山的空间再回顾一下，一个三维的点线面的构成，很人工，很建筑。但是她也是一面墙，一堵内部佛教世界与外部现实世界的墙，洞穴就是窗了。

作为意象的空间，她很建筑，但是墙的背后呢？还是墙，一个无尽厚度的墙，实心的山。

所以，龙门是另一种意义的建筑空间，是与传统意义的建筑相反的，我们称之为反建筑或负建筑。这当然也是我们丢掉相机后觉悟的一部分。好了，能否把这一反建筑的特点加以利用呢？我们关注的反正是建筑内部的影响力，外表的形态自然不那么重要了。重要的是内外情境与交通的关系。有一句歌德的话"它没有核心，也没有表皮，它是一个整体。"我们的建筑内部与外部应当是个整体，你无法把它们分开。我们的博物馆本身就是佛教思想的实体对话外部现实世界的结果。

We designing a building, we have to first consider the relations between the size and the external environment, e.g. how to surround the building and how to provide a "field". The meaning of walls is about privacy and border, and produces the feeling of a site. Of course, we don't have to only depend on the walls to produce the feeling as a site. Indeed, any thing that impacts on the space can do it. But walls are the most immediate and direct way.

And columns are the second option, followed by the entrance-exit relation, and the relations between layer space and passage. Most of the buildings, or the traditional ones for people to use, are similar. The windows and walls are tools separating the outer space from the inner space. But both spaces are empty. The difference lies in the reason why they are empty. Now let's have a look at the space on the west of Longmen. It is a three-dimensional structure composed of dots, lines and surfaces, looking quite artificial like a building. But she is nothing but a wall separating the Buddha world from the outside reality world. The caves are the windows. But what lies behind the walls? Again, the thing behind the wall is another wall with limitless width, or the hill. So Longmen is an architectural space in contrary to the traditional architecture which we call anti architecture or negative architecture. This understanding emerges after we throw away our cameras. Then how can we use the anti–architecture? Anyway, we focus on the influence of the inner space in the building. And the form of the look is not important. As Goethe said, "It has no core or skin. It is just a whole body". So the internal and external space of our Museum should be a whole body that cannot be separated. And the Museum is the result after the Buddhism ideology communicates with the reality world.

石窟的空间为实体的山上，内凹的空间，空间为附属.
The spaces of caves were chiseled in the mountain. The recess is attached.

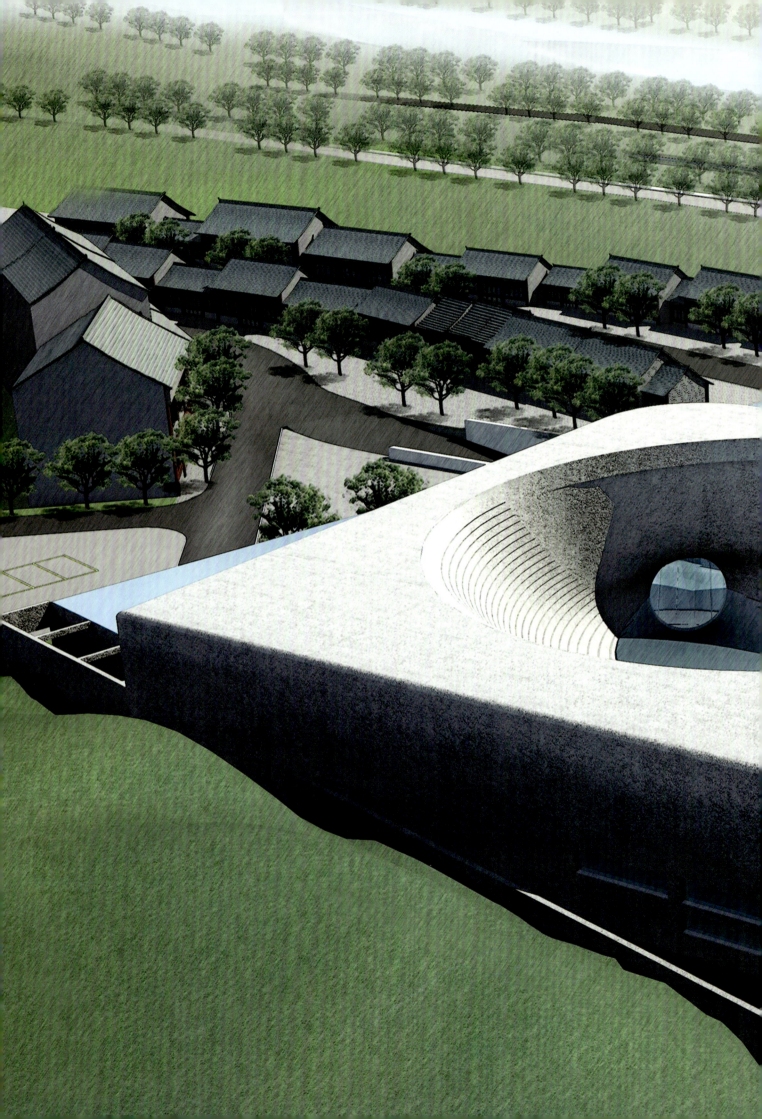

如果这一思想成立，我们就把反建筑的龙门变成空间特点，以内实外虚为方向来创作。何况我们还面临一个新的问题，从业主与龙门石窟管理局等各个方面来看，都希望减少建筑体量。原因很简单，这个地点距石窟太近，所有人都担心新建筑会破坏龙门石窟的景观，或给其带来过大的压力。这一点在投标阶段并没有强调，但我们自己也理解并赞同。所以决定将建筑大部分体量埋在地下，尽量少的暴露出来，同时结合建设用地依山面水的特点，再用堆土的方式隐藏建筑首层的部分外墙。本身建筑的思考方向就是浑然一体地去创造，又加之大部分隐藏的处理方式，就决定了我们对立面的放弃。

另外，在方案研究之初，老前辈蔡恒先生在与我聊天之时，提出这个主题很有趣，也很难做，不妨试试不要把它当建筑来做，这也给了我们很大的提示。刚开始时，我们无法克制内心的冲动，在表现力上大下功夫，完全背离我们之后思索与觉悟的方向，满脑子是对形式的执著，还是与肖老师等前辈高人的交流以后，我们才冷静下来。

反建筑的空间来源于龙门内实外构的特点，又贴合我们思索的现实世界与理想世界的过渡空间之理念，加之从山上开出来的巨石般的浑然一体的体量感，以及从内心里产生的它不是建筑的感觉……种种要素的沉淀后，我们与建筑最终可能的形态之间的距离越来越近了。

If this idea proved correct, we would change the anti-architecture into a form with full contents in the inside space and beautifully decorated arcade. And we were faced with another new question. Both the owner and the Longmen Grottoes Administration expressed their wish to reduce the size of the building. The reason is simple, it is too close to the Grottoes and many people worry that the new building would undermine the overall scene of the Longmen Grottoes or bring great pressure on it. This was not emphasized in the bidding process. But we understood and agreed with the idea. So our decision was to build most part of the Museum underground and only leave a small part above the ground. At the same time, we decided to hide some outer war of the Museum in the pile of mound. It means we have forsaken the arcade style.

Besides, in the pre-design preparations, the famous architect Cai Heng, when talking with us, pointed that this theme was quite interesting but much difficult. But he suggested we should have a trial that quit treating her like a building. His encouragement gave us great force. So in the beginning, we found it difficult to curb our inner impulse and made a lot of effort in the expression, which was far from the careful thinking and conclusion we made. But after exchanging opinions with Xiao, our mind full of persistence in the forms began to cool down.

The anti-architecture space comes from the special structure of the Longmen Grottoes, close to the idea of transitional space from the reality world to the ideal world. More importantly, the rock-like feeling produced from the suspended grottoes and carvings, and our inner feeling that she is not a building at all have steered us closer to the final form of the Museum.

从上至下：龙门石窟博物馆手绘图、鸟瞰手绘
图、人视手绘图、景观手绘图
From above to below
The hand-sketching of Longmen museum;
The hand-sketching of aerial view of Longmen museum;
The hand-sketching of human's view;
The hand-sketching of landscape;

(四) 佛教哲学的"空"和"圆"

(IV) The "emptiness" and "interpenetra-tion" of Buddhism philosophy

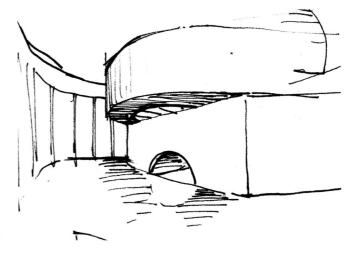

　　这段文字是很难写的，因为在短短的篇幅中写出佛教哲学的中心思想已经很难，何况还要阐明与建筑的关联。再有就是把佛教哲学空间化也是有无必要性的问题，在此只能试着分析了。在之前的几个理念的基础上，我们得到了创作的力量源泉、方向、性格特征、主题的定位。基本上得出模糊的答案，缺少的只是画龙点睛的一笔，也是最难的一笔了。

　　作为现代的建筑师，我们当然要与时俱进地看待建筑，前文的阐述是基于对世俗眼光于"符号"性的自觉继承之危险性的提醒。而当我们以"现代"为武器，并拒绝"符号"的时候，建筑的空间深层心理暗示，或影响力从何而来呢？当一个有经验的建筑师勤奋工作若干年后，他必定具备通过几何学原理，及对流线与功能的把握，并且得当地运用材料，创造合理及美观空间的能力。"普遍性"的空间评判标准是基本存在的，从中获取部分"特殊性"也不难，难的是空间的"场"的意义。"场所"是我们对空间的另一个解释，在德语中"场所"是"Ort"，其拉丁语的最初意思是"矛"的尖部，从这儿可以感觉到场所的感觉是有心理穿透力的。我们对场所的打造是要在普遍性的基础上，尽量具备与使用者内心需求相对话的影响力，特别是文化与纪念性建筑。

　　龙门博物馆是以龙门石窟为背景的佛教艺术博物馆。在这里，我们对空间影响力的创造，还是要回到佛教哲学中来找寻。佛教源于印度，但发扬光大在中国，特别是"禅宗"主导地位的确立，标志着佛教的中国化。谈到这里，产生两个命题，一是佛教哲学的核心思想；二是由传统佛教至"禅宗"的变化的根源。谈到所谓核心思想，没有哪一个宗教可以几句话概括的。我在做方案之初看了很多佛教的书籍，寻找一个标准答案，结果是越看越糊涂。哲学家与思想家或学者们有一个通病，不喜欢直说，总是绕着说。无非两个目的，一是显得内容难以捉摸，深不可测；二是回避说错的风险。当然，的确很难说清，而我认为敢做就要敢当，何况理论的目的不是解释或说明，而是自己的理解与延伸，否则看经书就好了，废话干嘛。我在一段时间的了解后，感到虽各家颇有不同，很多共性的东西还是一致的，在此不一一描述了。我自己断章取义总结一下佛教吧。

　　佛教的哲学思想来源于释迦摩尼在菩提树下的觉悟。其后的经文均是由弟子记录的释迦摩尼的语录，并由之归纳总结的结果。所以佛教通常是以四个字开头"如是我闻"，以四个字结尾——"信受奉行"。"佛"的原意如前文所述，是"觉者"而非"神灵"，所以佛教严格来说是无神论的宗教，最多是个人崇拜。佛教全部的思想体系就是围绕着"觉悟"一词。到了"禅宗"强调"顿悟"，表面看略掉了过程，总让人觉得是偷懒，而实际上结果也是偷懒。但其初衷正相反，其强调的是过程，结果不要去强求。这是禅宗鼻祖大师们的高明与智慧的体现，也是对"形式"之执著的放弃，对心性本质之源的求索。其成为一代宗教大派是因为其传播，正如现代社会传媒的力量。而传媒是最靠不住的，所以传播固然是好的，但也出现不同的理解与派系，例如"禅宗"之中国。但本质就是本质。对我而言最重要的还是"觉悟"一词的理解。常常有鸿篇巨著，扬扬洒洒十几万字的描写，在我而言就是"解脱"二字，并且通过"般若"来完成解脱。佛教就是帮助人解脱的

Writing the paper brought me many pains because it is extremely difficult to discuss the essence of Buddhism philosophy, and the relationship between the Buddhism and architecture in a short paper like this. We can only try to analyze the possibility and necessity of transforming the Buddhism philosophy into the space. Based on the few ideas we have discussed, we gain the driving force, orientation, features and theme of this paper, though half explicitly. Now we need to make the finishing touch, which is also the most difficult part.

As the architects living in the modern times, we have to have a modern view of the architecture. The above comments are to remind us of the danger of spontaneously inheriting the commonly-recognized "symbol feature" of the architecture. However, when we are trying to refuse the "symbol" and claiming we are the modern architects, how can we produce the deep psychological implications, or how can we generate the influence of the architecture? An experienced architecture working in the field for years will certainly have the ability to create the rational and aesthetic space with the principles of geometry, the understanding of streamlines and functions and by properly employing the materials. There are always the "generic" criteria judging the space and it is not difficult to extract some particularity from the "generic" criteria. The most difficult part is to define the meaning of "site". We sometimes interpret the space as "site". In the language of German, the "site" means "ort", whose Latin origin means the "point of a spear". We can feel the site often produces the sense of psychological penetration. So when making a site, in particular a cultural or commemorating building, we must try the best to produce the power of influence that can intercourse with the inner needs of its users on the basis of considering the "generic features" of the architecture.

Longmen Museum is a Buddha art museum in the context of Longmen Grottoes. So how we create the influence on the space has to be decided by the essence of Buddha philosophy. Buddhism had its origin in India and became popular in China. The dominant role of Zen in the Chinese culture can best illustrate that Buddhism has been localized in China. And this fact will produce two topics, namely, the core ideology of Buddha philosophy, and why the traditional Buddhism evolved into the Zen. No religion can summarize its core ideology in a few words. After reviewing a number of Buddhism books at the beginning of my design work trying to find an answer to this problem, I was totally puzzled. The philosophers, thinkers and scholars share one thing in common, that is, they would never give the direct answer at all. There are two reasons for their doing so. First, such way can make their theories or comments seem profound and complex. Second, it can avoid the risk of making mistakes. I find it difficult to agree with such approach. The objective of a theory is not to explain or illustrate something. Instead, it helps people to understand

哲学，解脱也是人们皈依佛教的原因。佛教哲学体系的构成与思想分支的拓展当然也是有意义的，是宗教影响力及哲学涵盖面与纵深的探索，也是哲学存在的意义。"解脱"容易解释，解脱什么，与觉悟什么就难了，而佛教的博大精深就体现在这里。这里我写的不是佛教论文，就不去深表，欲知详情如何，且听下回分解吧。但是就广大受众而言，解脱的当然是心灵的枷锁，完成一种心灵或心性的升华。

"般若"是生命中的智慧的意思，是教人解脱的方法论，这种智慧我的总结就是"空"和"圆"，也是我们讨论的重点。因为"解脱"的讨论本身就是执迷。最终想通了，也说明了，干脆别建博物馆了，那么执著干什么，空地挺好。甲方也好，建筑师也好，不就都解脱了吗。像禅宗那样，偷个懒吧。

"空"是"缘起性空"。"圆"是"圆融"，圆融无碍，圆满无尽。"缘起性空"也有多个解释，也一样篇幅无限。我再断章取义一次吧，就是存在的一切诸法源自（无常无我）无实有的"空"性，无"形"无"相"，无自性，如幻如化。所谓"诸法不自生，亦不从他生"及"诸行无常，诸法无我"，就是"诸法无自性"。无常无我就是无自性，无自性就是空。白话一些说，就是世上的"法"，是看不见，摸不到的，即不能以本体（自我）为参照，亦不能以客体（外部世界）为参照，是一种存在，又不依任何而存在，又存在于一切之中。在与不在之间的就是空。法本身就是自然万物围绕生发的一种无法用语言或概念去分析的法则。万物自有其生发循环的途径，而法则局于其中，空无一片。这个"空"玄妙无比，又难以捉摸，很像老子的"道"。而"悟道"与"觉悟"也是类似的。佛教就是引领你去悟这个"空"。现在知道唐僧为何给猴子起名"悟空"了吧，也知道为何孙悟空那么痛苦了吧。

佛说"万法归于一心"，禅宗六祖慧能说"万法在诸人性中"。我理解归于"一心"的心，就是空无的核心，万法循环其中，心不为所动，心就是法。慧能的"人性"即心性，是禅宗对"空"的新发展，也是传播的必然结果，否则皈依之路就太难以寻觅了，太容易走错了。禅宗的宗旨是探寻"心性的本源"，至此，佛教才真正发扬光大。原因也很简单，人们能理解个大概了，有了"心"了。但"空"一直是核心思想，并由此引发了各个佛教的派别，但万变不离其宗。

下面来谈谈"圆"。"圆"可以说是佛教在中国发展过程中提出的新观念。如果说"性空"的法居于中，循环其外的就是万物生发轮回的过程了。"圆"是圆通，圆融，是般若这一生命智慧的性格，"融通淘汰"的形象化解释。我个人理解圆是中国人浪漫智慧的体现，渗透着自觉的艺术光辉。禅家称"妙悟圆觉"。有了"圆"的存在，完善了中国佛教哲学体系，使心性与直觉成为其两大要点。圆的涵义之广泛，隐喻之多，是其它形态不能比拟的。"圆"，最具象也最抽象，它可作为哲学与艺术的形象代码，也是对哲学与艺术（客体）的浑整的认知的标志。这种"圆"往往是一种暗示或象征的喻体，是一种直觉智性的形象的把握，能够将艺术感知到的和尚未感知到的东西都表现出来。"志悦神圆"是感知的高级直觉感悟状态。

另外，圆也是一种融通，融是融化，通是通达，"圆融无碍"是无限通达，融化一切的表现，化去的仍然是一切实有，是执着。这也

something and expand their understanding into some other arenas. However, after spending some time studying the Buddhism, I feel the different branches of scholars have something in common.

The philosophy of Buddhism comes from the enlightenment of Sakyamuni under the Buddha tree. Then his students wrote down the comments and explanations made by Sakyamuni, and gave meaning to them to form the Buddhism scriptures. So we can find most of the Buddhism scriptures start with "As I heard" and end with "I believe and follow it". As mentioned above, the original meaning of "Buddha" is "the enlightened" instead of the "holy spirit". Strictly, Buddhism is a religion of atheism with some cult of the individual at most. The ideology of Buddhism is centered on the "enlightenment". Zen highlights the "epiphany" which seems like laziness. And it seems like laziness. But it emphasizes on the process instead of the result. The wisdom of the Zen masters is best shown in this point. They have abandoned the persistence in the "form" and pursue the source of nature. The Buddhism becomes popular because it has been disseminated extensively just as the modern media does. But too much dissemination has resulted in the emergence of many schools and understanding of Buddhism, such as Zen in China. But the nature is always the nature. For me, the most important thing is the understanding of "enlightenment" though there have been many thick books explaining it. I understand the "enlightenment" as the "relief", which is accomplished through "panna"(highest wisdom). In nature, Buddhism is a religion helping people with their "relief", which is often the pursuit of the Buddha believers. Of course, the Buddhism philosophical structure and the development of its schools are also meaningful because the different Buddhism schools are the explorations of the religious influence and the depth of Buddha philosophy. It is easy to explain what the "relief" is. But it is difficult to tell what is "relieved" or "enlightened". The difficulty shows the Buddhism is extensive and profound. Because this is not a paper of Buddhism, I will not spend too much time giving more explanations. For the common people, the shackles of mind are relieved and their spirit is sublimated.

Panna, meaning "highest wisdom", teaches us how to get the relief. Such wisdom, to be specific, is concerned about "emptiness" and "interpenetration", which is the focus point of our discussions. Indeed, the discussion of "relief" is also a kind of "obsession". If we have got relief in everything, it will not be necessary to build the Museum. In that case, nothing is meaningful, including the site, the owner and architects. Well, let's learn from the Zen and have a rest.

The "emptiness" refers to the "emptiness in nature" and "interpenetration" refers to the limitless and unhindered flowing feature. I will not waste time explaining the numerous understanding of "emptiness". Instead, I will focus on the commonly recognized understanding. "Emptiness" means that all existences are originated from the "empty nature". They are "formless", do not have their own "look" or their own nature. "All laws are not born automatically. And they are not born from others". It means that they don't have their own nature and this is explained as "emptiness". If we use the plain language, it means that all the "principles" or "laws" in the world are intangible. They are not able to consulted basing on the "identity" or "object". It is a kind of existence depending on nothing and existing in everything. The space between the existence and the inexistence is the "emptiness". The laws cannot be explained by words or concepts. But laws are everywhere. All things in the world have their own way of existing and cycling, in which we can always find the laws. And the "emptiness" in the laws is mysterious, difficult to touch and understand like the Tao of Lao Tzu. "Understanding the Tao" is quite similar to "enlightenment". The Buddhism is to steer you to understand the "emptiness" and that's why the Tang Monk would name his monkey student as Wukong (meaning "understanding the emptiness"). And we know now why Wukong would feel the great pains.

The Buddha said, "All laws end in the heart". The Zen Master Huineng also said, "All laws lie in human nature". Huineng, when talking about the "human nature", referred to the "nature of mind". His explanation developed the concept of "emptiness", which is also the result of dissemination to avoid the wrong orientation of the Buddhism followers. The purpose of Zen is to explore the "origin or source of the mind nature". And such concept has made the Buddhism widely accepted by people because they can have a rough understanding of the "emptiness". The "emptiness" is always the core idea of Buddhism, which has generated all schools of Buddhism.

Now let's talk about the "interpenetration", which was then a new idea proposed by the Chinese Buddhism. The "understanding of emptiness" focuses on laws. But there is always a process in which all things are born and recycled. The "interpenetration" represents the personality of panna. As I understand, the interpenetration is also the representation of the Chinese romance and wisdom that has its art meaning. Zen also calls it "skilled understanding and enlightened interpenetration". The emergence of the concept of "interpenetration" has improved the Chinese Buddha philosophy and highlighted the two points of mind nature and intuition. Indeed, the "interpenetration" boasts many connotations and metaphors. It is most specific and abstract at the same time. It can also be used as the image code of philosophy and art, symbolizing the overall cognition of the art (object) by philosophy. Such "interpenetration" is often the object of suggestion or symbol, and the image of the instinct wisdom. It can demonstrate things perceived or not

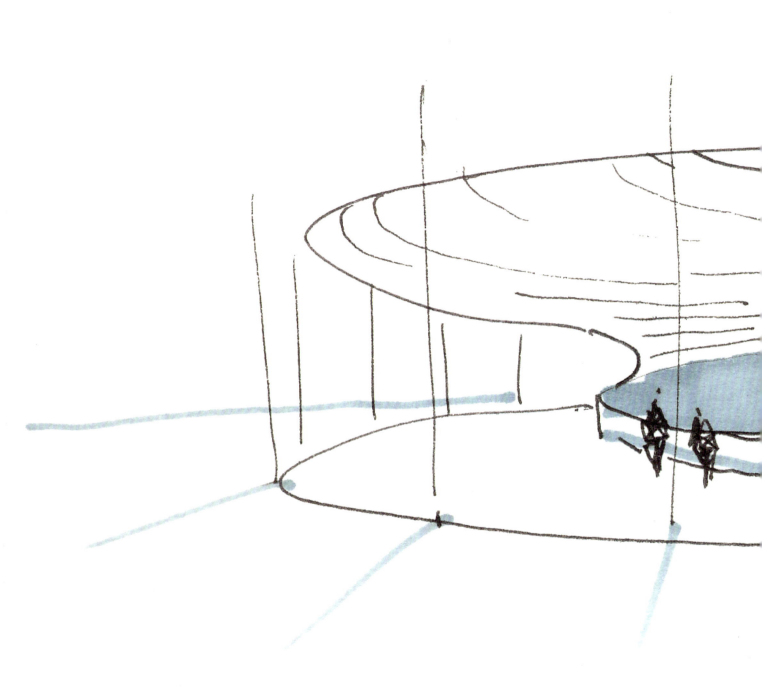

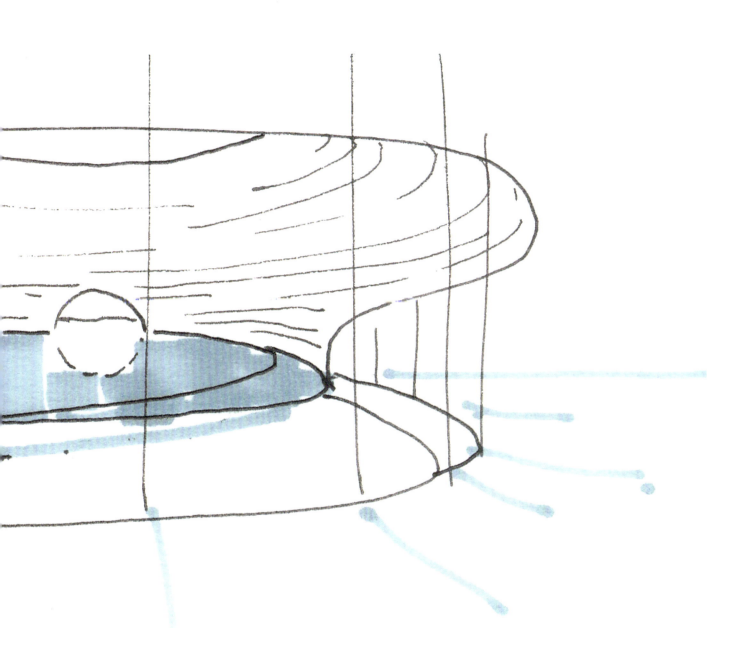

体现出中国文化的一种包容性的特质。因为圆融，我们融通一切，包括外来的，敌对的。这也是中国文明长久延续的秘诀。"包容就是智慧，智慧就是包容"。圆也代表圆满，完美，主伴俱足，浑然一体，是辩证法中的和谐的代表。例如道家的"太极"所表现的阴阳调和。故"圆满无尽"的完美状态，成为佛教以至道家共有的，甚至结合了"儒"家的"和"的观念，变成中国文化中最高境界的和谐圆满的意象图景。

"空"与"圆"的结合就完整地表达出佛教哲学的，特别是中国佛教哲学的"般若"的智慧，也就是中国佛教在哲学层面上的理解了。而佛教实际的表象又体现出另一种景象，出现了各种菩萨，各种神，如五百罗汉等等。这当然是佛教的在百姓层面传播过程中，被神化了的部分。当然，这些诸神们都是原本的"觉悟"的人。朴实的人民在寻求解脱苦难与憧憬美好来生的过程中，将其神化为心灵的寄托，同时佛教也利用了人们的这一心理，打出"普渡众生"的旗号，以利于其在世上的传播。并与"道""儒"相抗衡。

以上，我们将中国佛教的核心哲学思想用"空"与"圆"做了解释。现在就第二个层面，就佛教中国化的根源与禅宗的产生做一个补充阐述。因为这一点是与我们下面就龙门博物馆的设计问题有最为直接关系的。前文已就"心性"及"圆"的角度说明了佛教中国化的几个要点，这里我们重点讲的是"儒"与"道"对佛教产生的影响。中国文化体系是"儒、释、道"三教各自发展，相互影响，以至相互融合的过程的结果。孤立地看待任何一个都是不全面，不正确的。还是以佛教为例，虽然佛教有自成一体的理论基础，但是在传播过程中，其自身理论体系受到外部政治环境变化以及早已归为文化正统的"儒"学的挤压，并受到文人阶层对道家文化推崇信至的压力，不得不自我改革，扩展自身的理论，以适应环境的变化。禅宗就是在此基础上逐步产生并发展起来的，也就是所谓的"适者生存"吧。基于篇幅与时间和主题的限制，在此不能展开"儒、释、道"的大篇幅分析，只能重点将最主要的，也是我就禅宗的产生之最直接感悟的一点，来做分析。这就是禅宗在"缘起性空"的基础上，吸收了"道"家的辩证法思想，即"有无之境"，为其"性空"的理论作了有效的补充。特别是"无"。禅宗以"无"去悟道，使"悟空"成为感性的个体体验和高层次的直觉感悟，在感悟自身中飞跃获"悟"。其领悟神明，神超理得的圆寂状态，便是获得了认识和精神的超越的审美感悟境界。并且将道家对"人"的关注变成佛教"心性与直觉观"的基础，以迎合文人阶层的关注。禅宗于"儒"家的吸收，来源于其"治世"思想中的"和"的观念，作为其"圆"的思想的补充。"圆"本身是圆融，圆满的和谐观，儒家的"和"将"圆"提升至社会与心理及伦理的和谐的层面。"圆满"成为深入人心的人生准则。而佛教的圆满也由此变成佛教世俗化的象征之一，也成为求佛拜佛的主题之一。相对佛教般若的智慧"圆"，这是一种观念上的倒退，但同时换来了广大信徒的爱戴。顿悟的基础其实也是道家的直觉哲学，此外还有对个体开悟的强烈专注。更重要的是，它同时还符合中国儒家的一种传统信念，即"人皆可以为尧舜"，每个人都可以开悟而超凡入圣，得到解脱。

perceived by the art.

Besides, interpenetration is the process of reaching everything and fusing everything. It disposes of all tangibles and obsessions. And it explains why the Chinese culture is inclusive. Thanks to interpenetration, we can accept everything, including the exotic and hostile things. This is the secret of the Chinese civilization lasting so many centuries. "The wisdom is inclusive and the inclusive is wisdom". The interpenetration also means completeness, perfection and integrity. It is the best example of harmony in the dialectics, such as the harmony between Yin and Yang in the Taiqi of Taoism. As a result, Buddhism and Taoism share the notion of "boundless interpenetration", which even incorporates the concept of "harmony" in Confucianism. It is the highest level of the Chinese culture, producing an image of perfection and harmony.

By combining the "emptiness" and "interpenetration", we can fully express the Buddha philosophy, in particular the panna wisdom in the Chinese Buddha philosophy. And the Buddhism presents all kinds Buddha and Gods, such as the 500 Arhats. Of course, these presentations are the results of the apotheosis when Buddhism was being disseminated among the common Chinese people. These gods were originally the "enlightened" people. The plain and simple people, while pursuing the relief of their sufferings and dreaming about their better next cycle of life, tended to believe in the existence of Buddha and gods. Taking advantage of such psychology, Buddhism claimed that they could take the common people to the paradise for broader dissemination of the religion and to compete with Taoism and Confucianism. Since we have explained the meaning of "emptiness" and "interpenetration" in the core Buddha philosophy, we now come to the second level and say something more about the localization of Buddhism in China and the birth of Zen because they are directly related to the design of Longmen Museum, which we will discuss later. In the previous part, we have illustrated several key points about the localized Buddhism in China from the perspective of "mind nature" and "interpenetration". Now we will talk something about the influence of "Confucianism" and "Taoism" on Buddhism. Indeed, the system of the Chinese culture is featured by the independently development of "Confucianism" "Taoism" and "Buddhism". But they also impact on and combine with one another. So studying only one of them will produce biased and false results. Let's take the Buddhism as an example. Though it has own theoretical foundation, the Buddhism had to find ways to reform itself and expand the theories because of the changes of the political environment and the challenge from the orthodox Confucianism and the educated people's advocating of Taoism. Based on this situation, Zen was born and developed, quite similar to the theory of "survival of the fittest". Due to the time and length limit of the paper, I will not give much

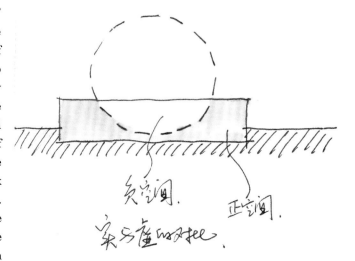

就龙门博物馆的设计，佛教哲学体系的"空"与"圆"将如何与其对话呢？这就面临一个问题，我们是把空与圆形象化看待，翻译在建筑空间中，还是将其安置在空间影响力的暗示层面上。或者在其哲学高度上，引用造型艺术的"形而上"的手段，将情境建筑的思想融入设计中，将"灵性"注入空间，以艺术意境的建构来指导创作。这就是前文所说的"画龙点睛"的那一笔了。"空"和"圆"的结合，是佛教般若的智慧，而智慧的体现就需要意境了。

"空与圆"的哲学内涵本身就有空间的意味，"内空外实，内虚外构，圆融无碍，圆满无尽"等等都是可以放入空间的形象思维中来的。建筑作为一种空间的艺术，可以避开结构逻辑的约束，成为直接感动心灵的力量。空间是灵动的，空间的本质虽然抓不住，摸不到，可是这种虚无，即无形，又无量，其大无外，其小无内，可供心灵驰骋。非常类似佛教的"性空"的思想。而"圆"本身就是空间常用手段之一，它的抽象性与饱满度给场所感的形成提供了最便捷的条件。圆形另一个特点就是有始有终或无始无终。每一个点即是起点也是终点，是循环，轮回的直接象征。而将圆进化成"球"，三维的圆，就形成了空间的量，抽象的可实可虚的"量"。可以承载无限的内容，也同时可以虚空一切，与外界形成"有无，虚实"的对比。内部只留下情境的灵动了。

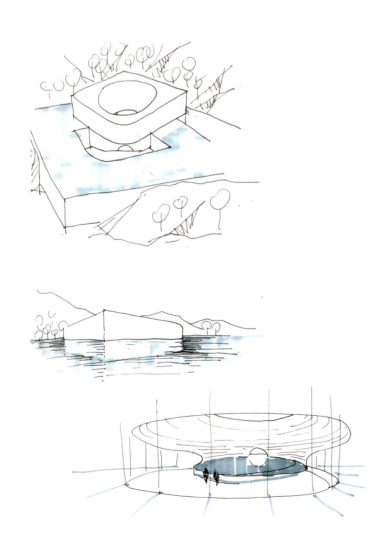

discussion about the "Confucianism", "Taoism" and "Buddhism". Instead, I will focus on analyzing the birth of Zen, which gives me the greatest enlightenment. Based on the "empty nature", Zen has absorbed some dialectic thought of Taoism, namely, "the level of everything and nothing", to supplement its "empty nature" theory. In particular, the Zen uses the theory of "nothing" in the effort to understand the "Tao", making the "understanding of emptiness" the direct and high-level conception of the individuals. The state of harmony silence meaning the understanding of ultimate laws, is the level of aesthetic conception exceeding the cognition and spirit. Besides, the Zen also employed the Taoism's focus on "humans" as the foundation of its "mind nature and intuition" to cater to the needs of the educated group. And Zen absorbed the concept of "harmony" in the "governance" idea of Confucianism as a supplement to its "interpenetration", which in itself is an idea of harmony. The "harmony" in the Confucianism has helped upgrade the "interpenetration" to the psychological and ethic level. And "completeness" becomes the well-recognized life philosophy. As a result, the completeness (an aspect of interpenetration) in Buddhism has become one of the symbols of worldly Buddhism. It is also one of the pursuits for the people who worship Buddha. Compared with the panna wisdom, it is some kind of retrogression. But the reward is the support of more believers. The foundation of epiphany is the instinct philosophy of Taoism and the strong attention of mass individuals. More importantly, it conforms to the traditional notion in the Confucianism, that is, "everyone can become a sacred person".

Then as far as the design of Longmen Museum, how can the "emptiness" and "interpenetration" in the Buddhism philosophy talk with it? We face a problem of whether to visualize the "emptiness" and "interpenetration" and interpret them in the buildings, or just leave it a hint of the spatial influence. Or should we use the "above the form" means to integrate the idea of context building into our design so that the "smartness" is placed in the space? This is the crucial point mentioned above. The combination of "emptiness" and "interpenetration" shows the panna wisdom, which needs the context.

The philosophy of "emptiness" and "interpenetration" in itself contains the meaning of space. We can consider the "empty interior and full exterior, boundless interpenetration". As the art of space, the building can avoid the constraint of structural logics and become the power that directly touches the mind. The space is smart. Though its nature is intangible, the intangible feature also allows the freedom of mind, which is quite similar of "empty nature" in Buddhism. And the "circle" is one of the tools used in space, whose abstract and full nature provides the best opportunity to establish the feeling of site. The other feature of a circle is that it has starts and ends, or no ends or starts. Each point can be the starting point or ending point. It is the direct symbol of cycle. When the circle evolves into a three-dimensional "ball", the amount of the space is formed, which is abstract, tangible or intangible. The "ball" can hold limitless contents and empty everything, forming the sharp contrast with the exterior space. So only the smartness of the context is left within the Museum.

（五）"心性和直觉"的情境空间
(V) The context space of "mind nature and intuition"

"心性和直觉"前文已叙述过,是中国佛教哲学思想的两大要点,它同时也是佛教融合"道家"与"儒家"哲学思想后的产物,是印度佛教传统刻板的经院式逻辑的一种反动。而"心性和直觉"也可以说是中国艺术领域中不变的主题。佛教的"空与圆"所倡导的"性空"的世界,与道家的"有无""虚实"的辩证思想相结合,构成了中国式的"意境圆"的象征体。

"中国意境的灵性在于虚空中显示出流动,似乎有一种流动的音乐——生命的律动,气韵的鼓荡,万象的纷呈节奏(一阴一阳,一虚一实)。这种超时空的生命节奏感,音乐感构成了中国意境圆的象征体的生命特质。"——姜耕玉《艺术辩证法》。

从这段话来看,中国意境的建筑可以最好地诠释"建筑是凝固的音乐"。当然,这里说的中国意境的建筑主要是指园林建筑了。中国的园林艺术就是这样一种呈现诗画的艺术。建筑居于园林之中,并不因为其建筑本身的功能而存在,而是造景的道具。且中国园林不能脱离诗画而存在,中国古典艺术中的"韵""神""趣""意""境""气"等范畴,均能理解为"有意味的形式"的子范畴,而它们每每又体现了中国艺术诗性智慧的独特迷离的色彩。所有的手段,哲学、艺术、诗性的运用,无非是要感动人。中国人对"人"的关注是"儒、释、道"三教合一的基础。也是有别于西方基督教的强势存在,多教并行,融合发展的秘诀。最典型的例子就是《西游记》了。一趟貌似千辛万苦的取经之路,变成了游戏于三教世界中的人性的温情之旅。

反观建筑的本质,我们再来寻求"情境"世界的关怀吧。"与人类纤细、柔弱的体质相比,建筑物显得异常的结实与长寿,甚至会让人感觉到它似乎在嘲笑人类短暂的生命,这就使人越发讨厌建筑物,可是建筑物的这种时刻张扬的不可逆转性实在令人厌恶。"——隈研吾。我想这都是因为现代建筑领域中,大家过多的对技术的关注,反之对情感关注的太少吧。这也关乎尺度,大体量的建筑对人的压迫感相对小尺度的园林建筑自然是大得多的。而都市建设中往往以大型公共建筑的体量来衡量权势与财力,建筑成为了炫耀实力的道具。而建筑师们在竞争与生存的压力下,不断的迎合着业主的需求,一方面为图降低成本,以及提高所谓空间使用率,不惜以降低空间质量为代价创造庞大而无趣的压迫。

另一方面,他们害怕被说成只看业主脸色的御用建筑师,在惯性思维的控制下,将原来无趣的空间再加上不置可否的无聊的主题,玷污建筑物这一本质自由的存在。最后,他们发现真正能施展的空间只剩下外表皮了。于是建筑师们堕落成了包装设计者,甚至还不如,因为他们忘记了建筑本身令人讨厌的几个要素,消耗最大的财富与能源,来建造的,招摇的,长久占据空间与视野,并不可逆转的恶化生存环境的强势的罪魁祸首。

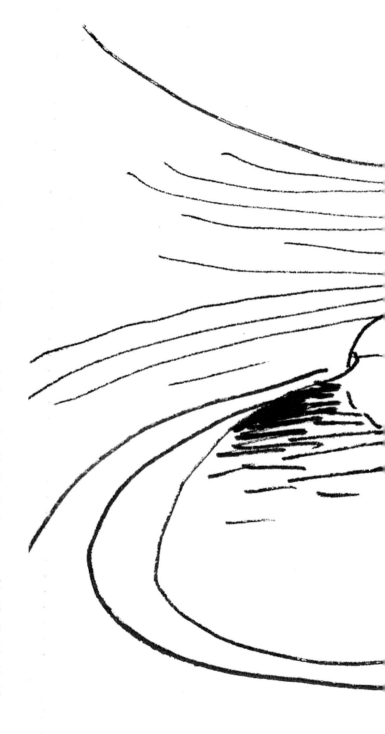

We have discussed the "mind nature and intuition" already. They are two key points of the Chinese Buddha philosophy and the result produced by the combination of "Buddhism", "Confucianism" and "Taoism". It goes against the rigid Indian Buddhism and the orthodox religious logic. The "mind nature and intuition" are the constant theme of the Chinese art. So the "empty nature" world advocated by the "emptiness and interpenetration" of Buddhism, when combined with the dialectic thought of Taoism featured by "everything and nothing" and "tangible and intangible", forms the Chinese symbol of "context interpenetration".

The Chinese context is smart because it demonstrates the flow in the "emptiness". It seems to produce the flowing music – the rhythm of life, the exciting atmosphere and the presentation of all phenomena (Yin and Yang, tangible and intangible). The rhythm of life and music that exceeds the time and space constitutes the life nature in the symbol of the Chinese context, the circle. – by Jiang Gengyu in his *Dialectics of Art*.

From his comments, we can find that buildings in the Chinese context can best explain that "the architecture is the solid music". Of course, the buildings in the Chinese context mentioned here mainly refer to the buildings in the Chinese landscape garden, which is the art with the poetic and painting features. The buildings in the landscape garden are used as part of the scene. Besides, the Chinese landscape gardens heavily depend on the poets and paintings. And the "atmosphere" "soul" "fun" "meaning", and "context" can all be understood as the sub-field of the "meaningful form", giving the unique and mysterious flavor to the Chinese art with poetic wisdom. The Chinese people use various methods, philosophy, art and poets just to move and touch others. And the Chinese focus on "human" is the foundation of combining "Confucianism, Zen and Taoism". It is also the reason why the Chinese religions and beliefs can co-exist with the powerful Christian region in the Western world. The most typical example is shown in the Chinese classic novel *The Pilgrimage to the West*. The difficult journey for the Buddhism scriptures was depicted as a heart-warming journey full of humanity in the reality world.

After studying the nature of architecture, let's have a look at the care from the "context" world. "Compared with the slim and gentle body of humans, the buildings are often sturdy and lasting. Sometimes we even feel they are mocking at us for our short life. And we begin to hate buildings. But they deserve the disgust because of the irreversible showoff ."– by Kengo Kuma. I believe that's because we focus more on the technologies and less on the emotions of the modern buildings. And it is also related to the size of buildings. The large buildings often produce greater pressure on people than the relatively small landscape buildings. In the metropolitans, the size of large public buildings is often the measurement of power and wealth. They keep catering to the needs of the owners and become the tool of showoff. On the one hand, the architects, facing fierce competition for survival, reduce the spatial quality of buildings to create the huge and meaningless pressure. On the other hand, they are also afraid of being labeled as the dependent architects only following the order of the owners. So their solution is just to find some dull themes for their buildings, thus staining the freedom of the buildings. Finally, they become the package designers because they find only the surface of the building is their work. They have forgotten the factors making the buildings disgusting, namely, consuming the maximum wealth and energy, occupying people's view for a long time in an aggressive way, and damaging the living environment irreversibly.

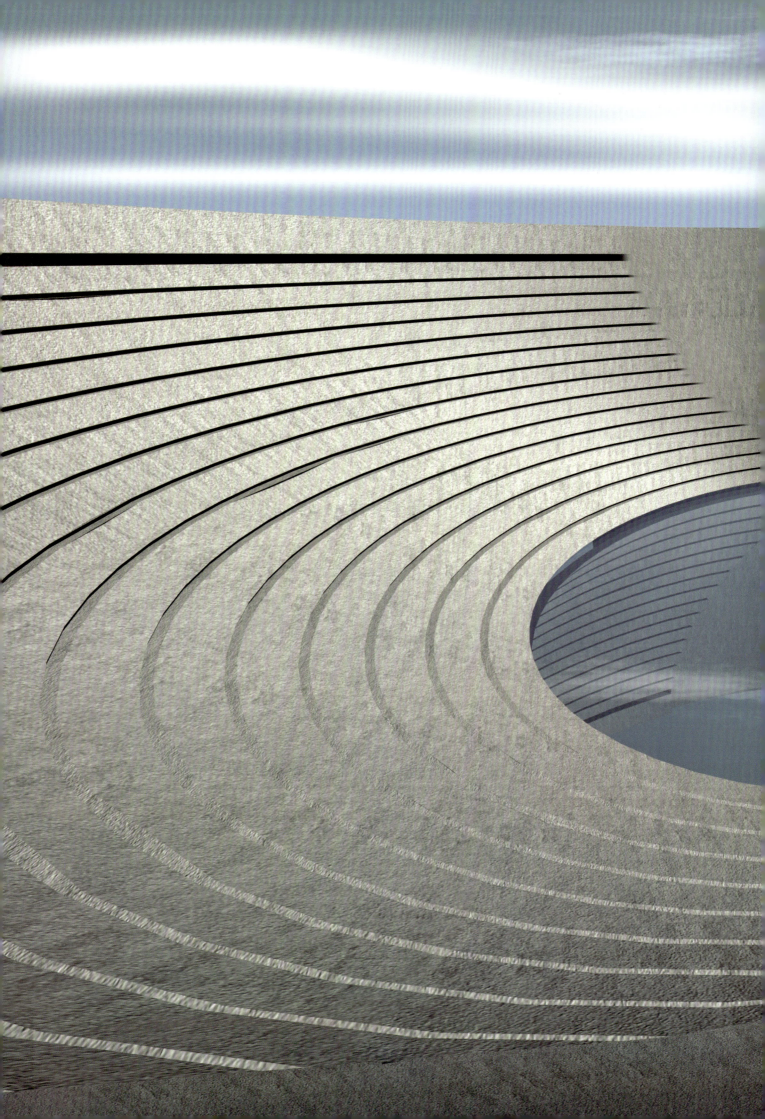

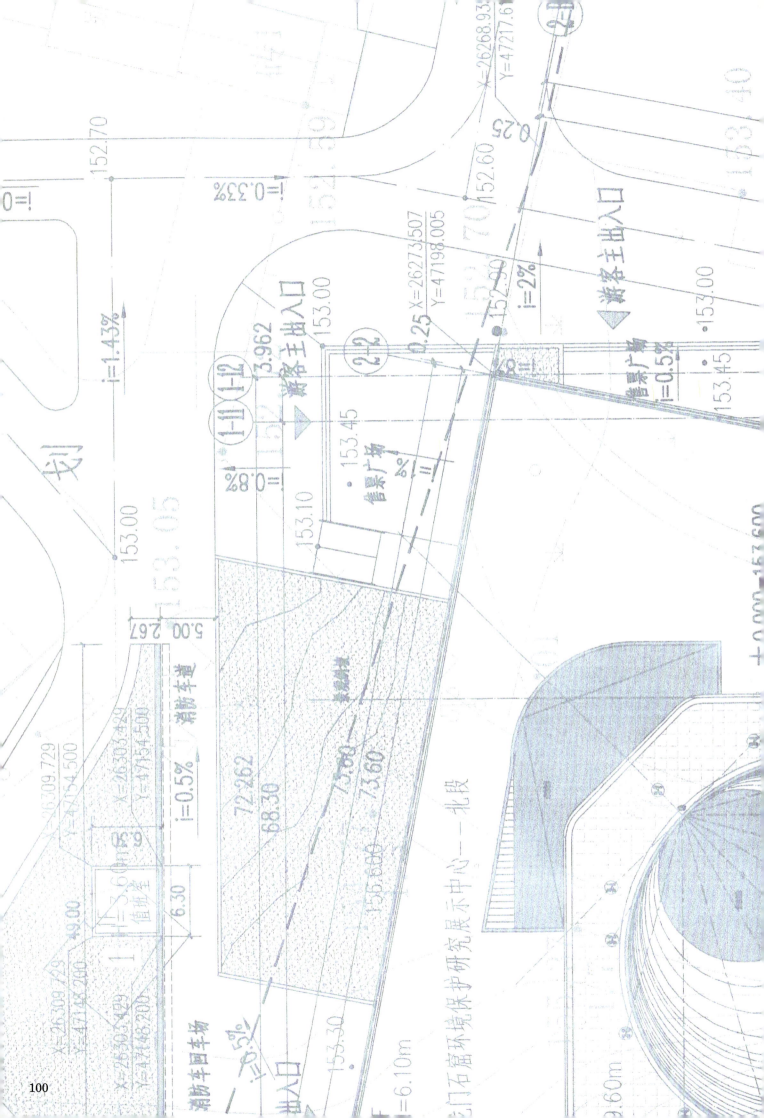

我写这些文字时，坐在老挝首都万象市中心街边的咖啡馆里，周围目之所及全都是不超过三层的小建筑，咖啡与街道保持着宜人的尺度。我可能要在这里改变这个城市的尺度感了。新的大型建筑会给这个城市带来现代与时尚的气息，也会打破城市中原有的平静与安详。而我好像除了提高它的亲和力与空间质量外，不能改变什么。建筑师在不得不选择为时代的主流欲望服务，同时还必须通过对这一现象的批判来改变这种"不觉悟"，哪怕改变是极其微妙的。佛教的哲学思想其实始终在反对的就是这种虚无的执著，在完成龙门博物馆的设计后，我才真正领悟了自己从事的事业的荒诞之处。也是基于此处，才想通过这次介绍龙门博物馆的机会，把自己的感悟拿出来，让人去品评与反思。

龙门博物馆的方案之路是艰难与漫长的，因为它与生俱来带着过多的包袱。历史的长河，佛教的哲学，洛阳的新亮点等等。本来就矛盾的形式与自由的关系倒向了形式的一边，建筑师无力与之抗衡。然而真正地完成它以后，那有限的自由反而背着无限的象征性被衬托得无比的可贵。感谢上苍对我们的眷顾，给予了我们如此的鼓舞，这有限的自由就是"情境"。一种在哲学与艺术的缝隙中存在的"心性与直觉"，一种建筑的灵性。

When writing the paper, I was sitting in a café in Vientiane City, the capital of Laos. All the buildings beside me were no more than 3 floors high. The café was standing in a proper distance from the street. Sitting in the café, I thought of changing the shape of the city. The new large buildings can bring the modern and fashion air to the city. But they will also break the peace and tranquility. As an architect, I can do nothing but improving its affinity and spatial quality. While serving the mainstream lust of their times, the architects have to do something to change the "none enlightenment" by criticizing the lust in a subtle way. Indeed, the Buddha philosophy is always opposing the "vain" persistence. After finalizing the design of Longmen Museum, I understand the absurdity of my work. That's why I share my understanding for you to review and comment since I have been given the opportunity to say something about the Longmen Museum.

The design of Longmen Museum took us a long time and many pains because of too many burdens on it. The history, the Buddha philosophy and the expectation of making it a new spotlight in Luoyang were all burdens on us. Now, we are biased for the form, which is often in contradiction with the freedom. And as architects, we don't have enough power to resist against the tilting. But after the Museum is completed, the limited freedom seems valuable because it contains limitless symbols. Thanks to the help of the Heaven, we got enough encouragement from the limited freedom, which is also deemed as the "context", or the "mind nature and intuition" existing in the chink between the philosophy and art as the spirit of architecture.

因为龙门博物馆是独立的建筑，地块面积也不大，我们不能将中国园林中的情境主义用于群落组合与递进的层进关系。真正能发挥的是我们通过观察与体悟从龙门"现实世界与理想世界的过渡空间"及"反建筑物的空间"中得出的以内示外，内实外虚的创造路线。这个内部的精神世界将成为所有矛盾与冲突的焦点。幸运的是，我们研究的对象正是"圆融"的高手。它亲自教会了我们"般若"的智慧，又给我们留出了思考的空间。看来佛祖他老人家在设计上是很专业的。告诉你该怎么观察，又不明言该怎么做，暗示你一堆谜语，却不告诉你标准答案，你硬着头皮报出答案，他也只是"拈花微笑"。你只能对自己产生怀疑，重新审视自己，寻找被误解为"嘲笑"的原因。若干次之后才明白没有对与错，只有"执"与"迷"。无论曾经认为自己多么正确，却只是在牛角尖里转。外面"空"的无边无际，只是自己看不到罢了。经过了一级一级的修炼，我们看到第七级浮屠了，只是还不知道中间还有几层隔着，再看看佛祖，他还只是微笑。

　　既然空间是整体的，我们把"情境"用在何处，才能"点睛"呢？

　　当我们关注实在的"有"时，忘了还有虚的"空"与"无"。释迦摩尼的微笑中，就是在告诉我们悟"空"。中国的意境中往往关注的是"空"与"无"。正因为容器是"空"的，才能收藏东西，这样其拥有的可能性就非常的丰富。这就是"物莫不因其所有，用其所无"的简单而普通的道理，却又揭示了一种普遍的深刻的自然法则。所以当我们关注一个充满佛教思想的空间时，还是要用"空"来表达。有了"空"，什么都不表达，也就是什么都表达过了。于是我们决定在原本反建筑思想的"实"中，再出现一个"空"。博物馆的流线是环形的线性路线，贯穿由上至下的所有展厅空间，中间是一个圆形的空腔。

　　空腔的尺度是在满足展厅空间需要的基础上尽量扩大的，大到足以影响室内的人。一团"气"貌似很轻，却又是不能承受之轻。而为了保留有大面积的展厅，空腔上大下小，到地下二层缩到最小，但并非同心圆的缩放，而是以东北角为切点，逐渐释放西南方向的空间。空腔最上方直径33.7米，最下方直径15米。

　　我们希望"空"的空间内什么都没有，甚至几乎是看不到的，观众从远处到进门都不知其存在，环绕其观看展览的过程也始终有从中空中传过来的体量的压力，一种强大的虚无的存在。人们感受到"空"的张力。这种张力令感官在各方面产生摩擦。而情感与想象，在积压，直到步入地下二层，光线与空气忽然被打开。人们步入了空腔最下方的开放部位，领悟了空间的本质。而那里面只有一汪水和碗型的空间投射阳光与阴影之间神秘的弧线。除此之外，什么物质都没有。而点睛之笔终于要登场了。

如来坐像
Seated Tathagata Statue

Because the Longmen Museum is an independent building covering a relatively small plot, we are not able to employ the Situationalism of the Chinese landscape gardens and create the cluster and layers in the Museum. What we can use is the idea of "tangible interior and intangible exterior" obtained from the definitions that "it is a space of transition between the reality world and the ideal world" and that "it is an anti-architecture space". The inner spiritual world will be the focal point of all conflicts and contradictions. Luckily, the object of our study is the perfect model of "interpenetration", which has taught us the "panna" wisdom and left us enough space of thought. It seems that the Buddha is also a design specialist who tells you how to observe in an implicit way. He just shows us some riddles without telling us the answer. When we give the answer in a hurried way, he just gives us a "Buddha smile". So you feel doubtful and start to review your answer, trying to find why you are "smiled at". After many reviews, you will understand nothing is right or wrong, and there is only the "obsession". No matter how much you feel you are right, you are only confined in a circle and not able to see the boundless "emptiness" beyond the circle. We have made so many reviews and modifications, and it seemed that we had seen the seventh pagoda. But we never knew how many more pagodas are in between. We looked at the Buddha, he just kept smiling.

Since the space is complete, where should we place the "context" to make the "crucial point"?

When we focused our attention on the tangible things, we neglected the "emptiness" and "nothing". Sakyamuni, in his smiles, told us to understand the "emptiness". In the Chinese context, we tend to focus on "emptiness" and "nothing". The utensils can hold because they are "empty", having enough space to hold the diversified contents. That's why we say "the utensils are valuable not because of what they own, but because they do not own". So when we consider a space full of Buddhist thinking, we have to use the concept of "emptiness". If there is "emptiness", even if we do nothing to express something, it is already expressed. As a result, we decided to create "emptiness" in our anti-architecture thought. The Museum is a linear route running through all the exhibition rooms surrounding a circular hollow in the middle. The size of the hollow is as large as possible while leaving enough space for the exhibition halls. It is large enough to influence all the people in the Museum. The seemingly light aura field it produces is actually very heavy. To leave enough room for the exhibition halls, the higher end of the hollow is larger than its lower end in underground level 2. Its size changes from its northeastern corner to the southwestern corner. The diameter of the upper end is 33.7 meters and that of its lower end is 15 meters. We hope to see nothing in the "empty" space. The visitors, when just entering the Museum, will not feel the existence of such a hollow. But when they walk around the hollow visiting the halls, they can always feel the pressure coming from the hollow, which is an "intangible" existence. So they can feel the tension produced by the "emptiness". Such tension produces frictions of their sensations. The emotions and imaginations of the visitors are being piled up until they walk to underground level 2, where the light and air comes in suddenly. They are now in the open space in the lower end of the hollow, understanding the nature of space. But on that level there is just a shallow water pool and the mysterious curl reflected by the bowl-shaped space, cutting off the sunlight and shadow. And the crucial point is right in front of us.

在碗底，圆形切点部位的下方，水池半圆形的出口，一轮满月通过倒影呈现在人们面前。我们堆积已久的情感终于释放出来。圆融无碍，圆满无尽的空间早已等在那里。

"使一个人深深震撼颤栗的某种东西，突然以一种不可言语的准确和精细变得可见可闻……犹如电光突然闪亮，带着必然性，毫不犹豫地获得形式——根本不容我选择。"

——尼采

它既是建筑的一个空间，也是一件展品，更是一种浪漫与诗意的美。它不是形式的表现，它是功能，因为"美的东西是功能。"

——丹下健三

满月的倒影是个空间转换的点，连通着两个内庭。碗型空腔外的庭院也是被展厅围绕的，是整个博物馆最明亮与开朗的室内空间。也是参观流线中情绪节奏设计的一个部分。所以满月的洞也是两个庭院互动的点，只是外面的人窥豹一斑，不明所以，里面的人清明透彻，圆寂"顿悟"罢了。

如果说之前的理念是定位，是"万法"，那这一条理念就是"心"了。"万法归于一心"就是"心性与直觉"的结果。全部思维的过程，就是一种"融通淘汰"的过程，方案之初千头万绪，除了被业主的期望值与自身的成功欲压得喘不过气。过程中又权衡再三，踌躇不前，不断重构，否定再否定，从牛角尖跳进另一个牛角尖。最后消化了所有，去掉了所有，从"有"至"无"，反而什么都在了。想到一段时间前看到的一句话"开始的时候，什么都不会来，中间的时候什么都留不住，最后的时候，全部都在"。

方案的过程现在看来与禅宗的修习与精进是如此的一致，自己的"执"是最大的对手，战胜自己就是解脱，虽然这是个不容易的事情。现在当然早就开始做很多别的项目了，却发现还是一样的作茧自缚。好在龙门博物馆的设计像个镜子，没事照照，就会给自己提个醒。

"菩提本无树，明镜亦非台，本来无一物，何处惹尘埃"。

——慧能

设计之路永不平坦，走下去就好，皈依心中的理想，永不停息。"皈依并不在一个处所，皈依是在路上"。

这些文字是关于理念的阐述的，写给关心龙门博物馆的所有人。是6年多关于该建筑的情感的集合。这里的理念也许不适合于别的建筑，但无论是什么主题，好的理念都是不容易得到的，需要建立对这一主题的情感，才具备前提。但是正因为它不容易得到才使它变得珍贵。像圣–埃克苏佩里在《小王子》里写道的"使沙漠变得这样美丽的，是它在什么地方隐藏着一眼井。"

再有半年，博物馆就能开馆了。很期待那一天，也很怕那一天。因为自己"觉悟"的只不过是目前这一级浮屠罢了。距第七级还不知差几层呢！我们现在最多是《射雕英雄传》开始时的"江南七怪"，千万别自我感觉太良好，华山论剑之前，是个人都能一口气把他们吹飞了。开馆那天，再去龙门石窟卢舍那大佛前时，她还只是微笑。

邹迎晞
2011年4月2日
结于老挝万象市

The circular cutting point is right at the bottom of the bowl. Through the half-circular exit of the pool, the full moon presented itself in front of us. We released our emotions completely. The space with boundless interpenetration was already there waiting for us.

"Something shaking us greatly suddenly becomes so tangible in an unspeakable and meticulous way. It is like a flash of lightening with inevitability... I have no other choice"

By Friedrich Nietzsche

It is both an architecture space and an exhibit, producing romance and poetic beauty. It is not formal, it is functional, because "beautiful things are functional."– by Tange Kenzou.

The reflection of the full moon is the point where the space transition happens. It connects the two internal courts. The court beyond the bowl-shaped hollow is also surrounded by the exhibition halls. It is the most bright domestic space in the Museum and part of the emotional rhythm design in the visiting route. So the hole in the full moon provides the opportunity for the two courts to interact with each other. The people outside the Museum will not understand it. And only those who have entered the Museum can be "enlightened".

All the principles mentioned above are the "laws" giving orientation. But this principle is about the "heart". "All laws end in the heart". So it is the result of "mind nature and intuition". The process of designing the architecture was really difficult and too many ideas came into the mind. We faced great pressure from the expectations of the owner and from our strong aspiration to make a success. Besides, we hesitated many times about which option to choose. After numerous re-designs and denials, we went from one extreme point to another. But finally we forgot everything. When we moved from "everything" to "nothing", we find we have had everything. The process reminds me of a comment I read not long ago, "In the beginning, nothing would come. In the middle, nothing could be retained. In the end, everything is here".

The process of designing the architecture is just the same as learning and practicing the Zen. The "obsession" is our biggest enemy. By conquering ourselves, we can get the relief. It was really difficult to realize this. Of course, I am now working on other projects. But I still find myself falling into the same dilemma as I did in designing the Longmen Museum. But Longmen Museum functions like a mirror and keeps reminding me of the lesson I have learned from it.

"The Buddha tree is not a tree. The mirror table is not a table. Since nothing exists, why bothering ourselves?"

By Huineng

The architecture design is not an easy job. We have to keep following the path and our ideal. "Buddhism belief does not happen in a site. It happens on the way."

Some parts in the paper are about our ideas for all people involved in or caring for the Longmen Museum. The paper has pooled all our feelings and emotions in the past 6 years. The ideas and principles might not suit the other buildings. But good principles are always valuable because they can only be reached when we love our work. The principles are valuable because it took efforts and hard work to get them. Just as Saint Exupery said in his *Le Petit Prince*, "The well hidden somewhere makes the desert beautiful".

The Museum will be open in half a year and I do look forward to the day. But I am also upset about the upcoming day because my "enlightenment" confines me to my current pagoda, far from the highest level. At most, we are like the first teachers of the novel *Eagle-shooting* Heroes and should never be arrogant. In front the top martial masters in the novel, they are nobody. On the inaugural ceremony of the Museum, the Grand Locana Buddha will still be smiling.

By Zou Xingxi

on April 2, 2011 in Vientiane City of Laos

技术数据图
Technical data diagram

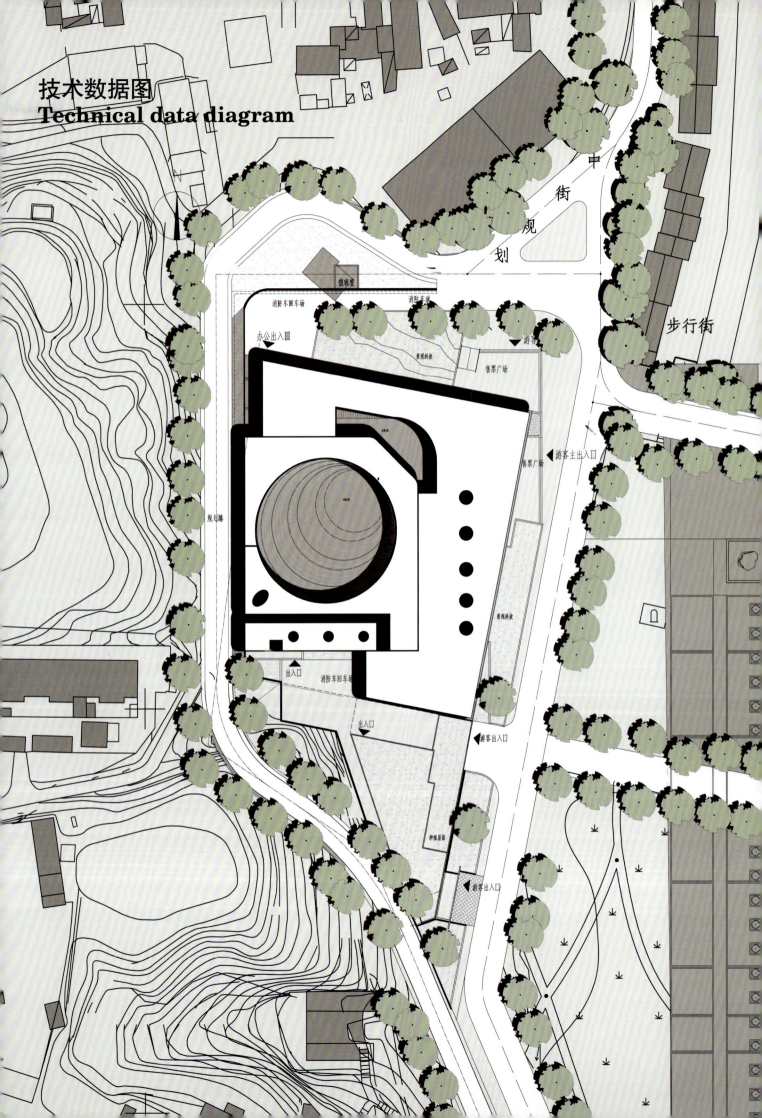

龙门总平面图
Longmen Master Plan

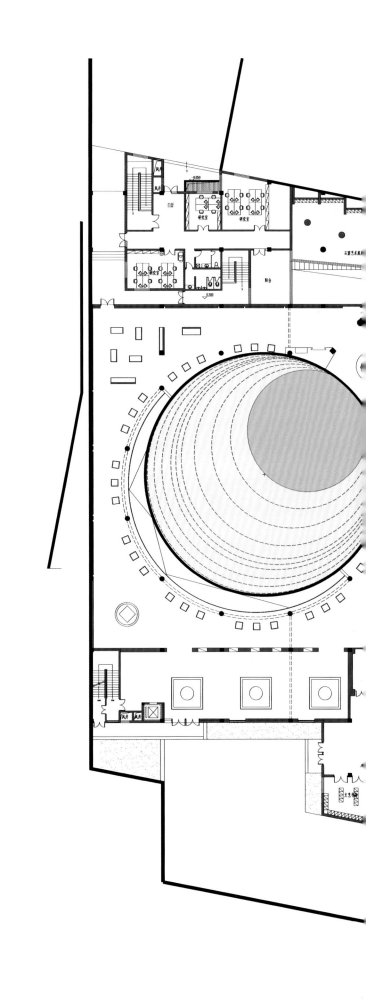

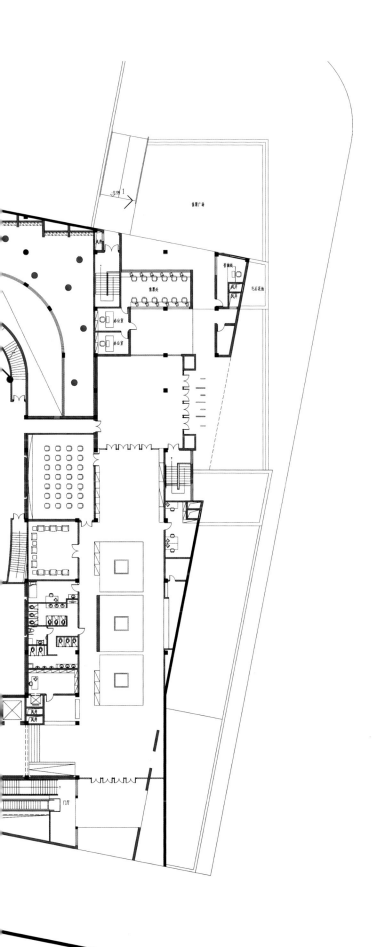

首层平面图 1:150

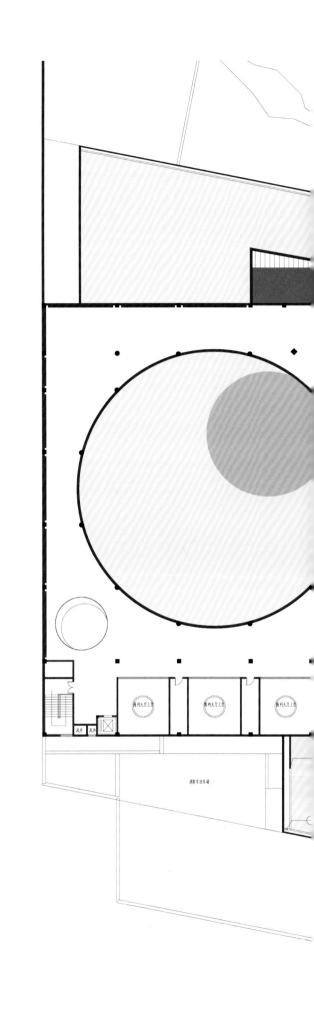

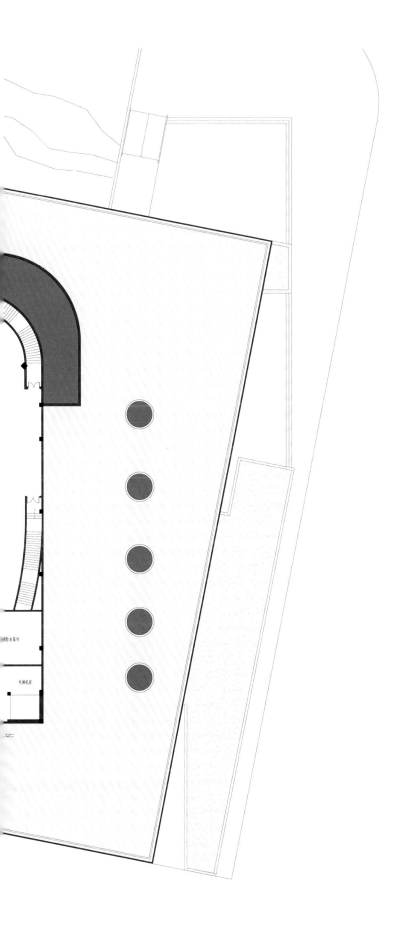

消防水泵房

电梯机房

二层平面图 1:150

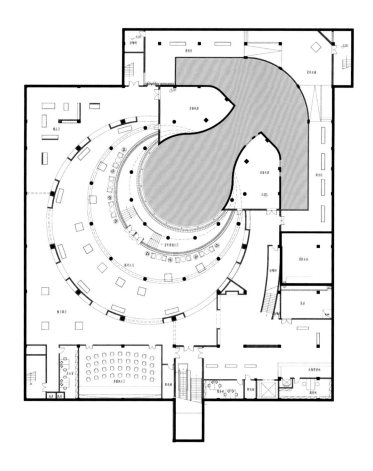

地下二层平面图 1:150

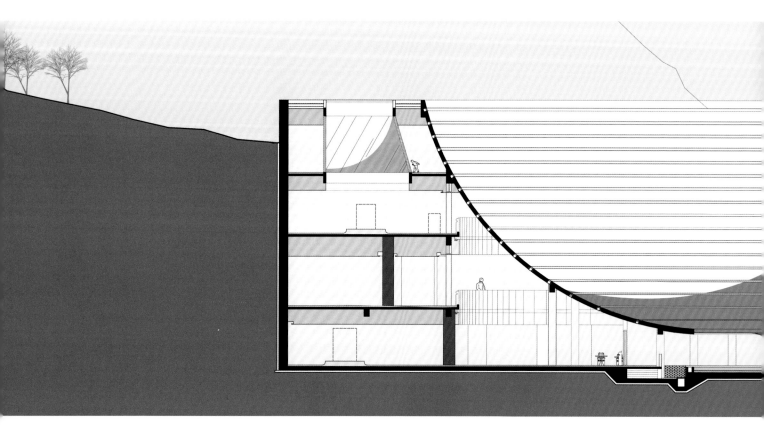

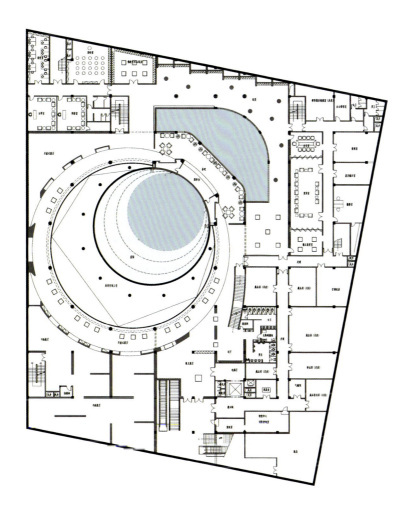

地下一层平面图 1:150

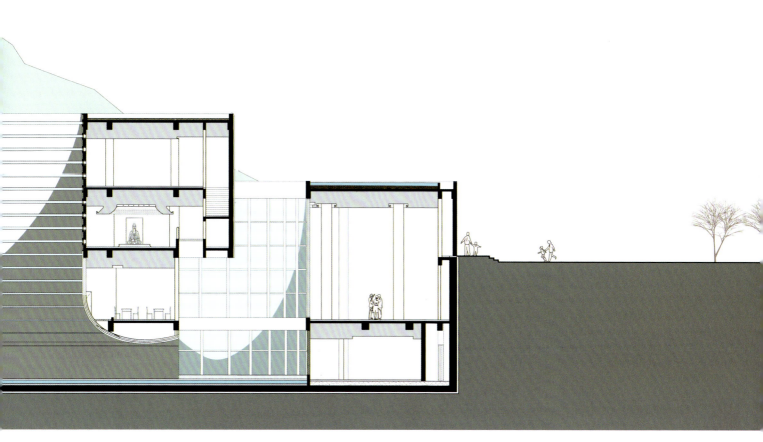

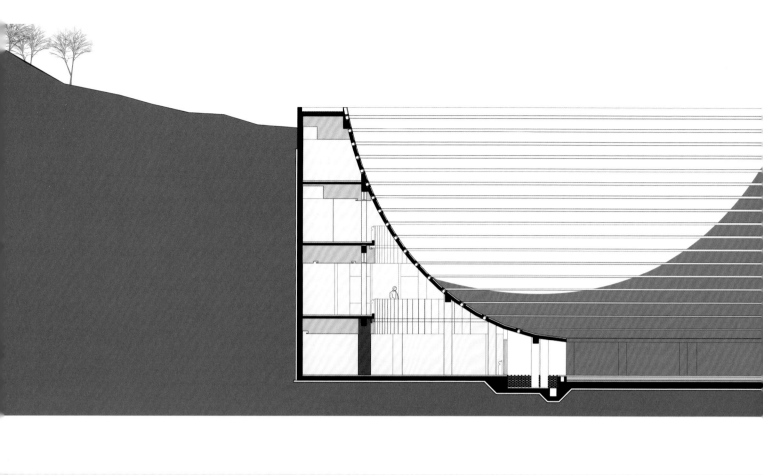

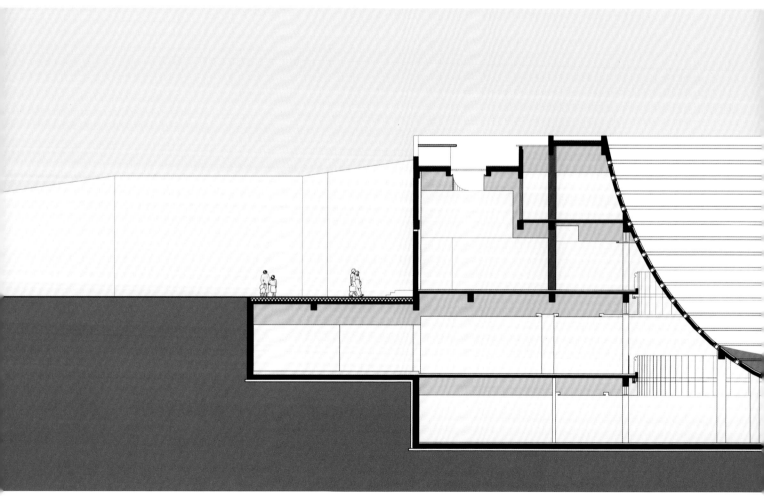

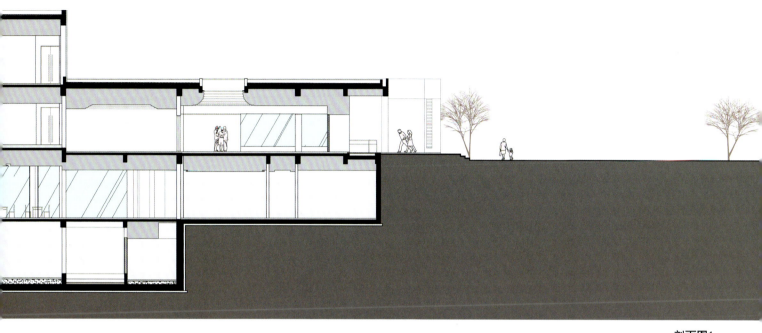

剖面图1
Section 1

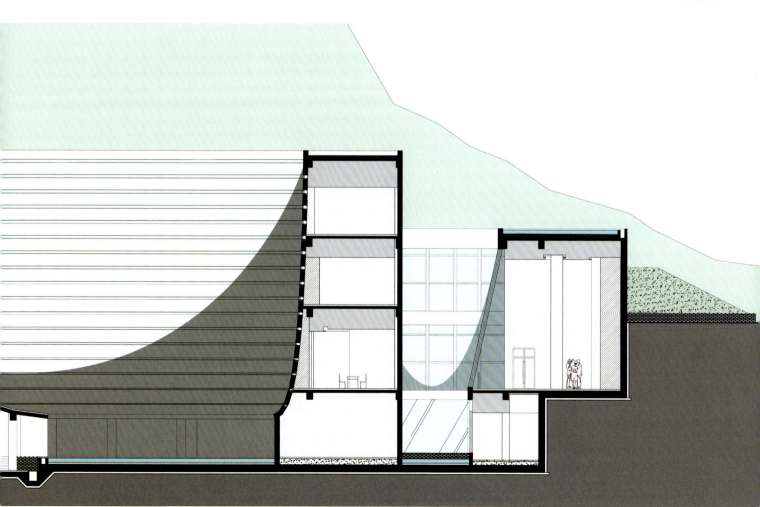

剖面图2
Section 2

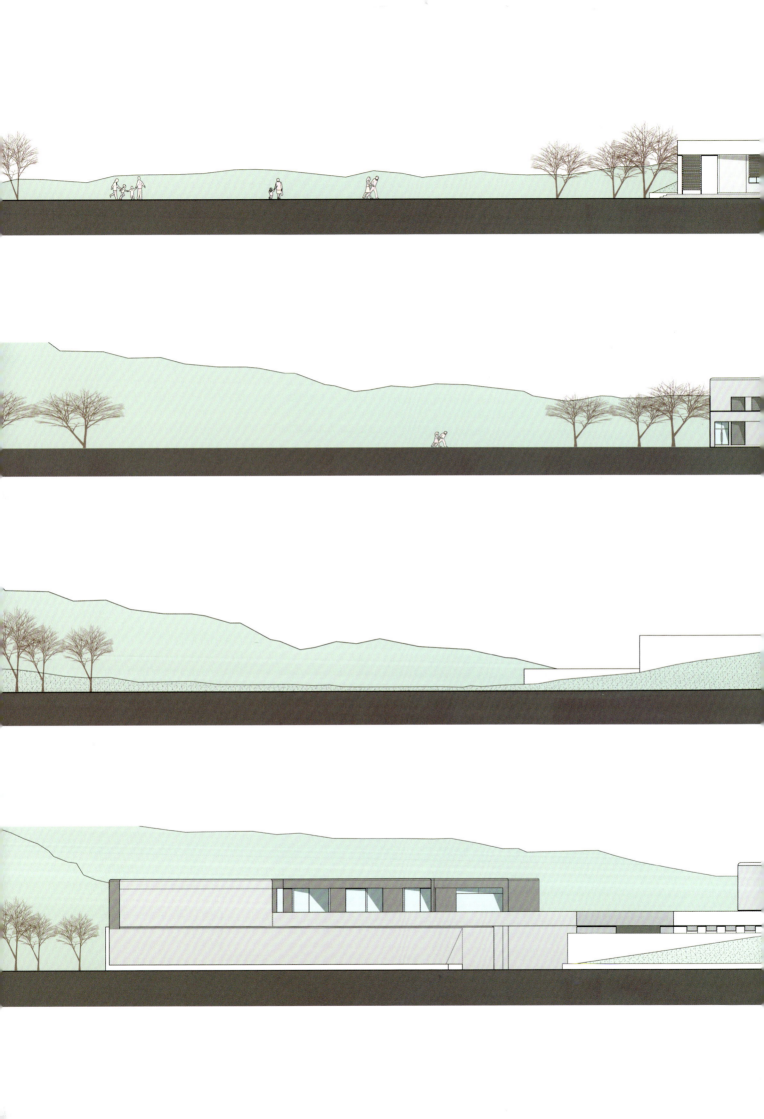

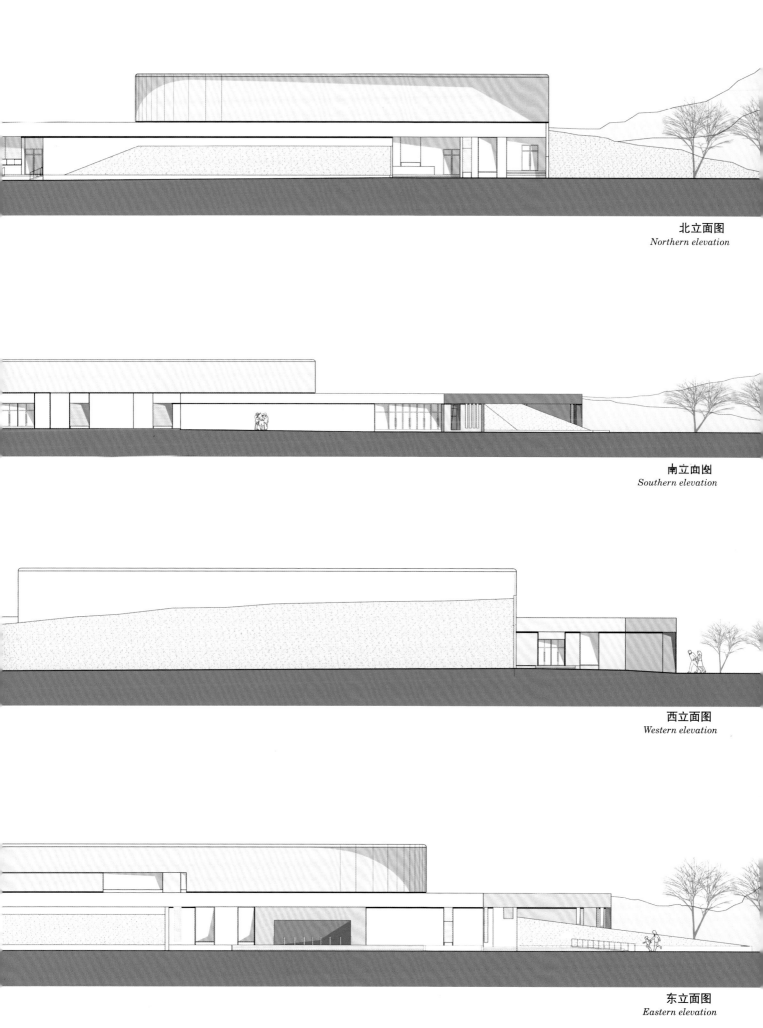

北立面图
Northern elevation

南立面图
Southern elevation

西立面图
Western elevation

东立面图
Eastern elevation

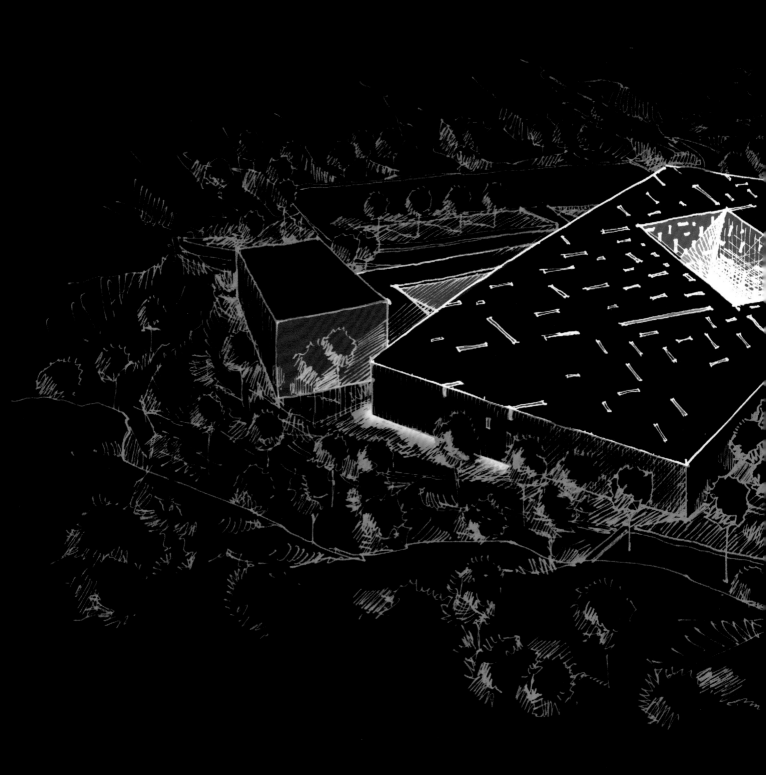

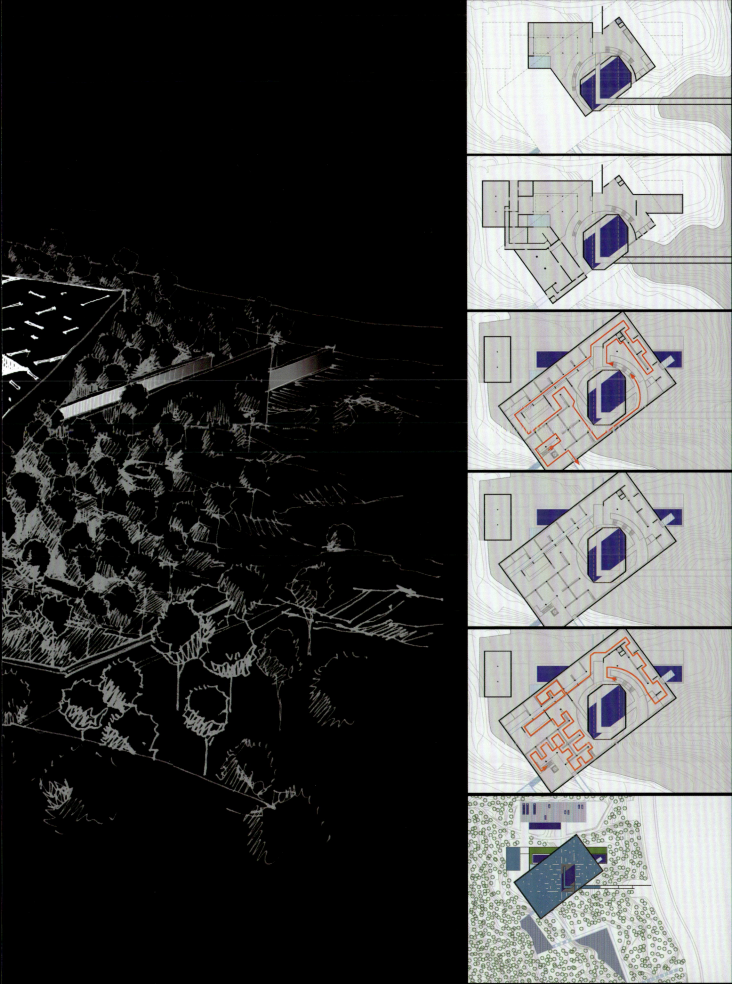

工作模型1
Analytic model 1

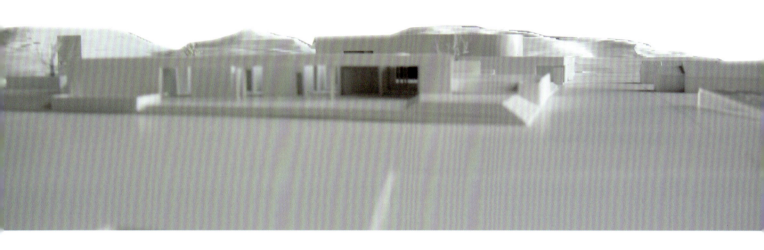

工作模型1
Analytic model 1

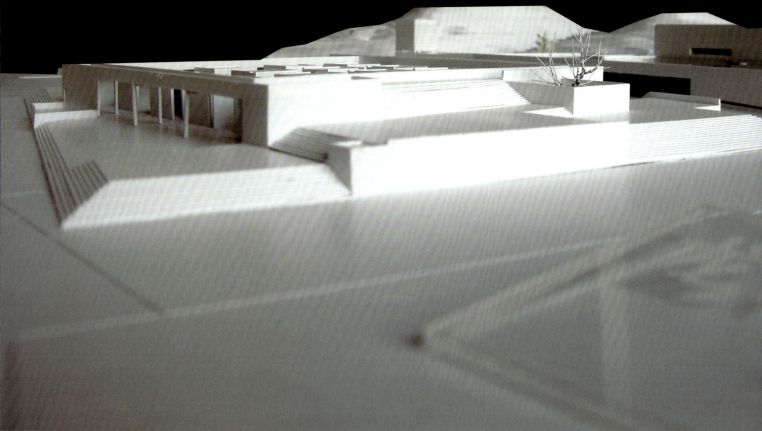

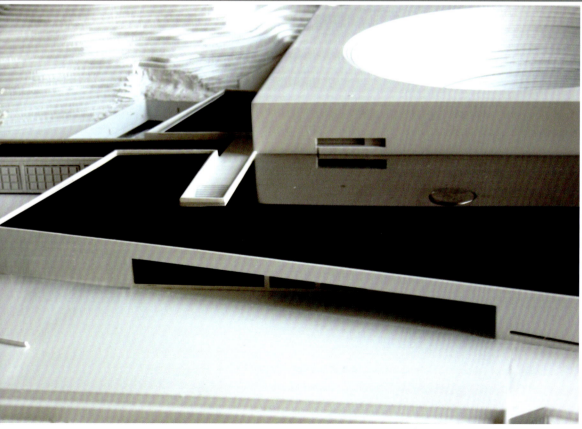

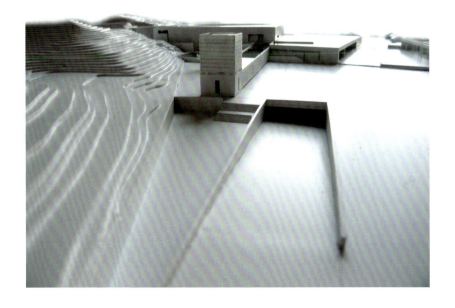

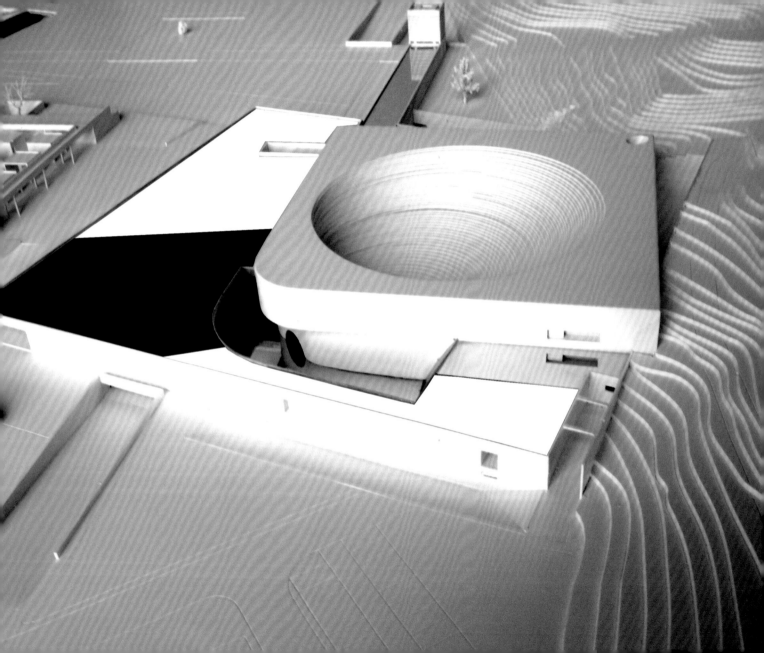

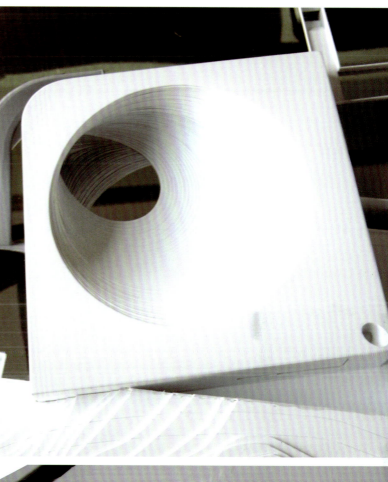

工作模型2
Analytic model 2

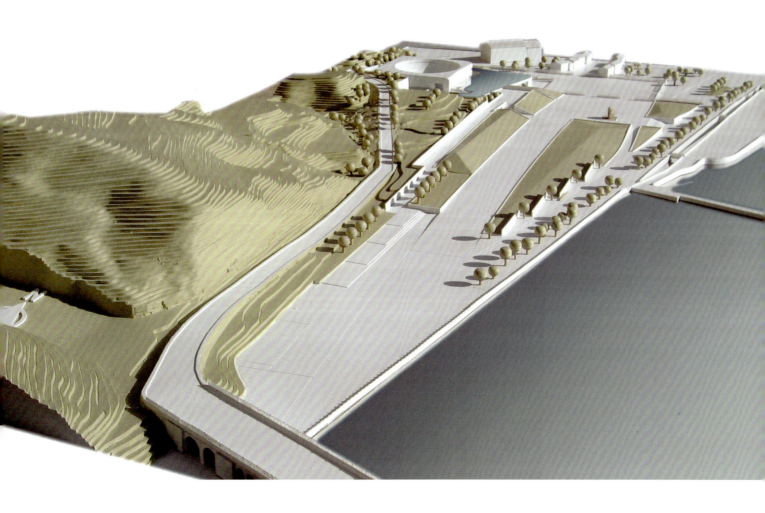

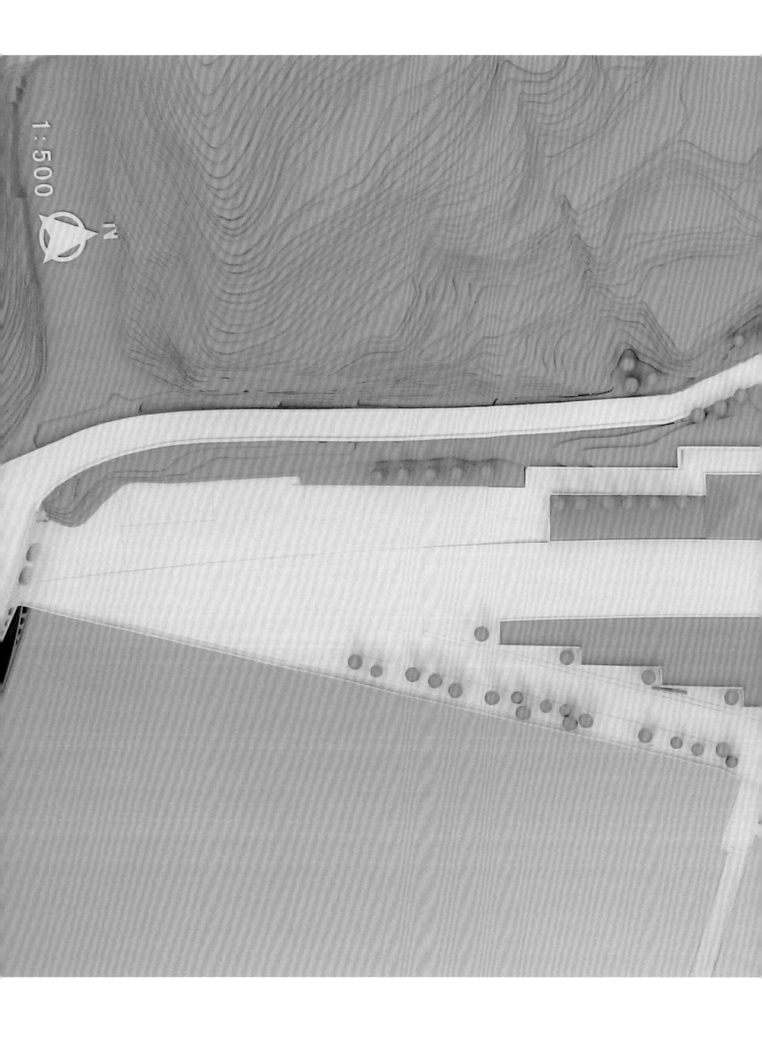

1:500

效果图
Architectural designsketch

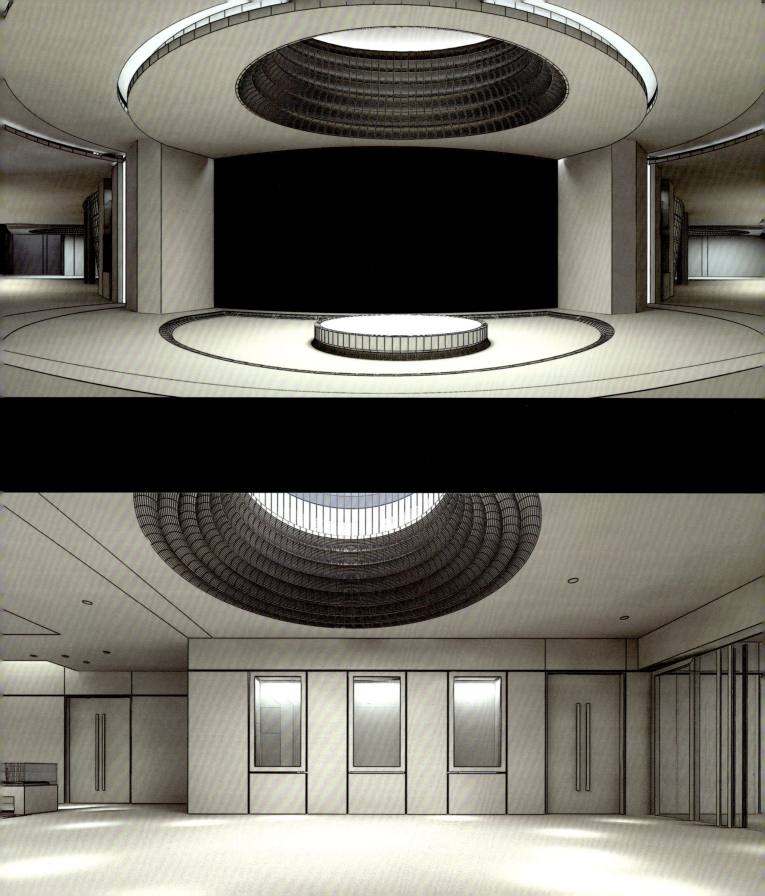

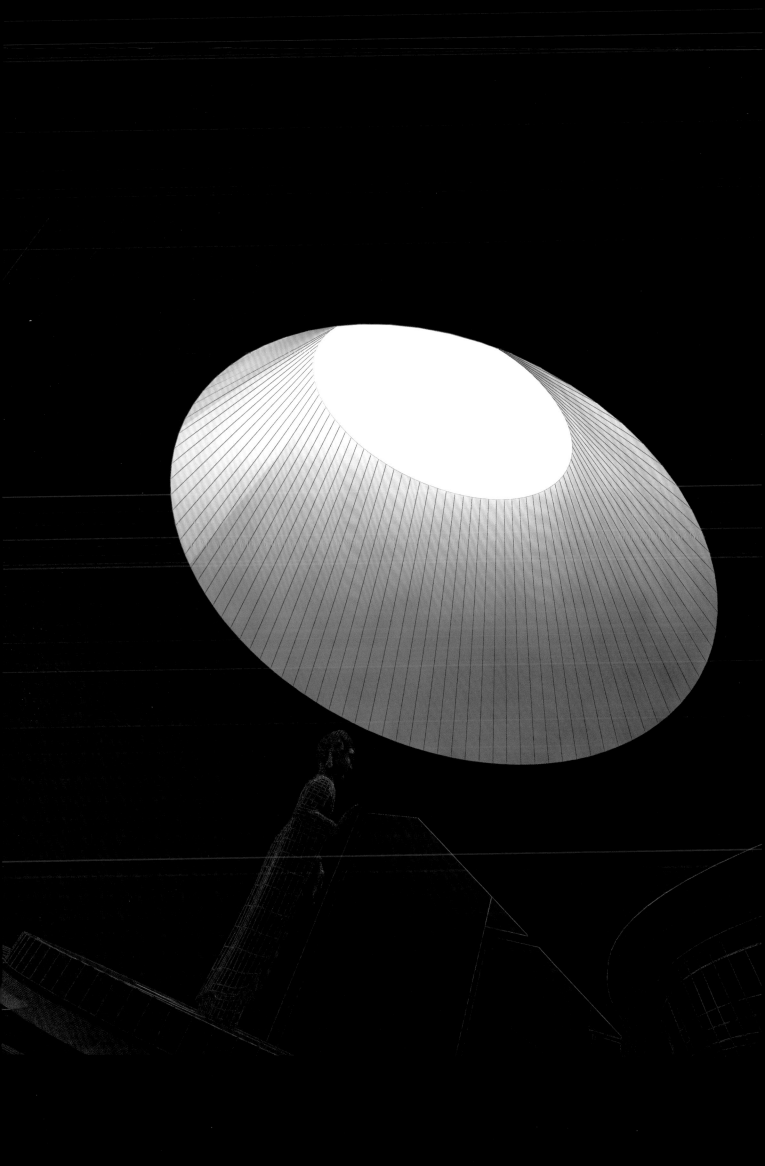

"有"与"无"（有无之境）

　　通过对龙门博物馆的设计理念的整理与阐述的重新过程，一方面整理了这几年对该项目的思绪，一方面也有了些新的感悟。特别是在写"空"的过程中，由佛教哲学上的"悟空"引出了中国艺术辩证的"有"与"无"。感到还是有必要在此题目下进一步分析，这也是自我拓展一下空间吧。利用这次机会，审视一下自己身上的"有"，而目的是探寻那未知的"无"。

　　中国人的艺术观与西方世界的艺术观很大的不同在于，西方运用抽象思辨的模式，中国运用直觉感悟的模式。抽象思辨属纯理论形态的思维，由整体出发，向微观发展，是一种自上而下，自大而小，由外及内的逻辑推理过程。例如丹纳《艺术哲学》，他提出"种族、环境、时代"三大因素对艺术品的本质面貌所产生的重要作用。书中详尽的分析艺术品诞生地的自然气候和精神"气候"，诸如宗教、政治、历史等等对艺术品产生的内在影响。而谢林的《艺术哲学》特别之处是提出"潜能"说，与亚里士多德的"潜能"说相呼应，是与黑格尔的《美学》共同对西方艺术哲学的重要补充，并着重于艺术辩证法的方向，但都着重于自上而下的思辨。而中国可以说正相反，我们反而是一种自下而上的直觉体验模式，也因此我们没有系统的古代艺术辩证法专著。从我们古代诗文、绘画、戏剧、舞蹈、书法、园林等的理论资料中，充满着艺术家们切入艺术本身的独到的辩证。例如"形神、虚实、曲直、方圆、有无、正反"等等，在阐释具体艺术辩证范畴时，总是深入艺术内部去发现和发掘自身体验到的独特的东西，以形象的语言方式表述其精要，并将抽象的理论形象化。中国古代文论，画论均表现出直观领悟高于推理思维的特征。由此可见，我们重感性，注重形象意会，重妙谛，而西方重本质，重规律，重比较，重价值的判断。但我们思维广度与深度可以空灵无际，却没有系统，相互重叠，且有疏漏。西方显示出理论的强大系统性的优势和力度，虽力主鸟瞰，高屋建瓴，但触及不到阴影后面，树荫之下那迷离与神秘之处，以及"只可意会，不可言传"的玄妙。在我看来这正是"有无之境"的迷人之处。"有无之境"表达的核心其实是科学思维的理性界限，与直觉体验无限伸展之间的辩证关系。抽象思辨所无力抵达的形象本体层，直觉体验却能够深入和企及。本文无意全面叙述艺术辩证的全貌，只想就"有无之境"的角度探索一种艺术与设计的方法。并且作为一名建筑师，对中国现代建筑艺术的现状与发展提出方向性的建议。

　　"物莫不因其所有，用其所无"，经典地表达出"有"与"无"的辩证关系。拿中国传统绘画为例，写意山水与花鸟画一直是中国文人们永恒不变的主题。我一直在想，为什么千百年以来这么多人画一个主题而画不厌呢？而且还有一种奇怪的现象，有人只画其中一种类型，例如画虎，画马，还有画驴的。在我最初的意识中，对国画是比较不屑的，因为的确有千百年来只画几张画的感觉，这几张画也是构图不同而已。直到有一天，在贵阳，进入我的同学家的客厅，看到墙上有一张小画，画的是一幅冬天的大树，枝叶全无，笔法简练，遒劲有力。一种莫名的感动突然涌上心头，让我的目光久久舍不得离去。画面留白很多，每条树枝以近乎奇怪的方式向四周发散，但放到一起又很均衡。单纯的黑白水墨，极简的画风，却表现出一种无以名状的丰富。这张画彻底的改变了我对国画的认识。画的作者是我这位贵阳同学的父亲，杨长槐。从此我开始关注中国

Reality vs Spirit

During summarizing and analyzing the design concept of Longmen Museum, I am trying to sort out some of my ideas towards the project in the past few years, meanwhile, there is some other new ideas arising in my mind. Especially during my composition, enlightened by the philosophical idea of "understanding the empty" in Buddhism, which leads to the Chinese artistic dialectical concepts of reality and spirit, It is necessary for me to give a further analysis to this topic. It is also the process of self-exploration, with the help of scrutiny of the reality of myself, to explore the unknown spirit.

The most obvious difference of artistic ideas between Chinese and western world is that westerners are more likely to adopt abstract model to analyzing the reality while Chinese apply their instinct and inspiration as a model. Abstract analyzing and thinking model is a kind of theoretical thinking, start with macro-world to explore micro-world. It is the process of logical deduce from higher to lower, from big to small, from outside to inside. For instance, Taine in his book "the Philosophy of Art" claimed that the three factors "race, environment and times" play a very important role in forming the essence and appearance of the works of art. He gave the detailed description and analysis in his book on how the natural climate and 'spiritual climate' such as religion, politics and history in the birthplace of works of art exert their influence over the inward of the works. The most unique viewpoint in Schelling's book is that he gave the concept of latent energy, corresponded with Aristotle's theory of potentiality. Together with the book "Aesthetics" written by Hegel, they are all very important supplement to western artistic philosophy. They all emphasis artistic dialectics, and followed the model of analyzing and thinking from top to bottom,contrary to traditional Chinese way of thinking, which emphasizes intuitional experience, characterized by from bottom to top. That is one of the reasons why there is no such monograph on artistic dialectics in ancient china's history. The recorded materials in ancient Chinese history such as poems and essays, drawings and paintings, dramas, dance, calligraphy, gardens show strong evidence of dialectics of artists. In a specific scope of artistic dialectics, they try to go deep inside the world of art to uncover characteristics by summarizing self-experiences and then to explain them with their vivid descriptions such as"the body and spirit, reality and fantasy, straight and curve, square and circle, being and not being, positive and negative"to visualize the abstractive theory. In ancient Chinese critic classics, theory of painting is characterized by intuition overweighing reasoned thinking. This shows that Chinese emphasize perceptual, visualizing and understanding by inside; pay much attention to tactics and knowhow, while westerners place particular stress on nature, regularity, comparison and value judgments. The range and depth of thoughts of Chinese is boundless yet not systematic, sometimes overlapping sometimes missing. The west's theory shows its powerful systematicness and strength. It gives a bird's eyes view or strategically situated overview, but can hardly touch the shadow-covered details—the mysteries and the wonders of things that as Chinese old saying goes,"not the words, but the meanings"or reading between the lines. From my point of view, this is the exact attractiveness of the theory of presence and absence. The core of the theory is to reveal the dialectical correlations' between rational limit of the scientific thinking and intuitional experience with its infinite extension. Abstract thinking cannot reach the visualized world while the intuitional experience can. This article cannot provide an overview to artistic dialectics, instead, from the viewpoint of "presence and absence theory", to explore a method for art and design. And as an architect, I would like to present my own suggestions to modern Chinese architecture and its development.

As the classical theory goes, "for all things, we make use of their property with their characters." This is the exact description of the dialectical correlation between reality and spirit. Take the traditional Chinese drawing and painting for example, landscape abbreviated in and flower-and-bird painting have always been the most favored themes for Chinese artists. I was wondering why such tradition lasts for thousands of years for so many people and never feel boring. Furthermore, it sounds strange for me that some people only focus on some specific animal, such as tiger, horse or even donkey. In my early age, I looked down on the traditional Chinese painting a little bit. For I thought, even with thousands of years of history, these paintings were almost the similar drawings, the only difference was the composition of such pictures.This idea kept unchanged until the day I visited my classmate's home in Guiyang. When I entered the living room of my classmate, I noticed a painting on the wall. It was a big tree in winter, no leaves but only branches. The style of drawing was concise but powerful and vigorous. I was totally absorbed and touched by this painting for quite a long time. There were lots of blanks in the painting, every branch dispersed in different directions in a strange way, but put it together, they are perfectly balanced. The simple black and white ink, the brief style of drawing presented me so many abundant meaningful things. It was the single painting that totally changed my understanding towards Chinese traditional painting.The painter was Yang Changkui, the father of my classmate in Guiyang. From then on, I began to focus on Chinese traditional paintings. Especially after reading the book "my quotation to Shi Tao's painting" written by

绘画，特别是看了吴冠中先生写的《我论石涛画语录》以后，才将心性进步到另一个我原先完全不了解，现在却深深着迷的新境界。中国画的秘密就是以"有"画"无"。我个人的看法，中国画画中的对象只是某种媒介，它传达的是画家对心境的流露。画面可以很小，对象可以单一，然而它表现出的情感与内涵却可以无限地放大，大到没有边际。从这一角度来看，无论画的是什么，都是不那么重要的了。所以中国画关注的是"无"，是相对存在于画面与构图中的"有"的，画面以外的东西。而笔墨的多少，笔法的功力，繁简的对比，主题的明暗都是为"无"而服务的。吴冠中先生说："笔墨等于零"，虽略显强硬，但于我而言却是至理名言。同时他还说过一句话，"艺术贵在无中生有"，又反过来在"有无之境"上做了补充。这里他强调的是主观因素的作用了，也是强调艺术家在动笔之前对对象赋予的情感，这个对象仍然是种媒介，只是是从画面外切入进来，体现在"有"中。同时还有一种偶然性在里面，因为结果是水墨韵染的必然，而水墨是很难控制的。毕加索有一句话："创作时如同从高处往下跳，头先着地或脚先着地，事先并无把握"。这也是就偶然性的描述。从偶然性里面也可以看出"有无"之间的微妙。吴老的话看似是说"有"，实际上还是在说"无"，"有"只是为"无"服务罢了。偶然性只是有无之间的一种调味品，令"有"相对"无"产生了些许的变化。毕加索只是放大了调味品的剂量，对结果充满新的期望罢了。因为他的作品是传统西方艺术哲学在穷途末路中的新曙光，也正是重视经验与直觉的开始。他决不是靠偶然性创作的人。

也许搞艺术创作的人经常有类似的感受，本来很满意的作品，为了增色，加上了新的修饰，结果反而破坏了之前的表现力，或变了味儿。这种情况在任何创作中都类似。那是因为"有"与"无"之间的平衡被改变的结果。中国人最擅长留有余地，经常有些描述，例如"话说一半儿，心领神会"，再如"犹把琵琶半遮面"等等，所以劲要使得有余地，"境与情"才有空间，艺术创作中也是如此。比如"留白"，中国画是很讲究留白的，然而画家着笔时脑子里想的不是画中剩余的空白，而是以空白开始画，空白处是"气"，是其与画外融通的关键所在。所谓的"疏可跑马，密不插针"是就构图的疏密而言的，而本质是"意在笔先""趣在法外"，早以"胸有成竹"了。而败笔就是所谓"画蛇添足"。

再用音乐来举例。人们欣赏音乐，总是评价旋律的优美，音色的幽扬。而旋律源自单个音符的组合方式。有趣的是，音乐也是有空间的，它是通过音符与音符之间的停顿来组成的，所以如果我们换个角度，音乐是在一个停顿到另一停顿之间调动人们的听觉的。一首乐曲也是由一开始的无声，渐进，序曲，高潮，结尾，再至另一个无声的过程。音符就是"有"，而停顿就是"无"。音乐本身就是"有无相生"的最典型的例子了。"大音希声"的含义就在于无声的间隙，包含了一切下一个可能。再比如诗歌，中国的古诗的魅力之一是精炼、概括，且意味深长，从不累赘。语言的感染力部分的类似音乐，也是停顿的艺术。而诗歌更极端，它以高度洗练的词句，创造深长的意境，引人遐想。词句的出现颇有画意，词句成为媒介，传达的是无言的情境。就是所谓"言有尽而意无穷"。刘熙载《艺概•诗概》中说"律诗之妙，全在无字处"。故言为"有"，意为"无"。诗歌亦是"有无相生"的佳例了。例如王维的"大漠孤烟直，长河落日圆"，一个"直"，一个"圆"除了表明形态外，暗示了一种广阔无垠，同时平和无风，清朗透彻中的苍凉、色彩、尺度及孤寂。中国的绝句无论"五言"或"七言"，有意地控制字数与押韵，反而在

Wu Guanzhong, the most famous Chinese painter, my thought was upgraded to a new stage. The stage I have not known before, but now totally amazed by it. The secret of Chinese painting is that it tries to convey the spirit through reality. According to my own understanding of Chinese painting, the objects in Chinese painting are only a kind of media; they reflect the painter's emotions. The picture may small in size, the object in the painting may be simple, but they can express the boundless ideas and emotions of the composers. From this point of view, the content in the painting does not count. So Chinese painting focuses on spirit rather than the objects in the painting. Here , the composition of the painting ,the style of drawing, the contrast between simplification and complexity , the light and shadow of the theme are all tools to serve the main idea of spirit. Mr Wu Guanzhong even claimed that "pen and ink are nothing", it seems too assertive, for me, I think it is the absolute truth. His another opinion" the value of art is it creates something out of nothing", is a supplement to the theory of presence and absence.Here he put much emphasis on subjective factors, that is, the feeling and emotions of the artists towards the objects they are trying to paint. Here again, the objects play as a media. They were drawn into the painting from the reality. There are some elements of contingency–for one can hardly control the dyeing process of the ink. Piccaso once said,a the process of drawing sometimes is just like jumping from the top, you can never make sure whether the head or feet hit the ground." This is the further explanation to contingency. Such contingency also reveals the subtle connections between reality and spirit. Mr Wu's opinion seems focus on reality; actually he lays stress on spirit. The reality serves as a tool to the reveal the spirit. Contingency here only plays as a favoring, to make a little bit change in reality when reflecting the main idea of spirit. Again, Piccaso only amplified his flavoring to make people full of expectation. His works are the new dawn to the traditional western philosophy of art which seems face a dead end. He started a new era of focusing on experience and intuition. He was never the person who merely rely on contingency.

Perhaps the following is the similar experience for many people who participate in creative jobs. The original perfect works were modified with several additional decorations; as a result, these decorations ruined the original expressiveness or changed the theme. That is because the balance between reality and spirit was changed or destroyed. The Chinese people are really good at leave some leeway. As the old saying goes,"only telling half of the story, to understand tacitly", "held the pipa partly to cover the face." Only enough blank space was kept can make a room for surroundings and emotions to grow. Chinese traditional paintings put much emphasis on void space or white space. Painters do not care too much to the void space when they begin to paint, they start with the void space, and the void space is a kind of ventilation hole to communicate with the surroundings. The density of the painting follows the following principle: it can be as spacious as a racecourse or dense as if you can't even find a place for a needle. The essence of painting is"conceiving before painting, and lingering charm beyond rules." That is, one knows all the answers before painting. A faulty part in a piece of painting is to ruin the painting by add something or paint the lily.

Music is another good example to illustrate the relationship between reality and spirit. People appreciate music. They give their comment to music such as graceful melody or sweet and elegant tune. Here again melody is composed by every single musical note. Interestingly, music too needs some space. The pause between musical notes composes music. In other word, music is a series of pauses between notes. A piece of music starts with silence, then advance gradually, over tune, climax, coda and finally, to another silence. Musical note here means reality, pause means spirit. Music itself is a typical example of the mixture of reality and spirit."The great sound is hard to hear" means pause plays very important role in music, pause means next possibility.

Poetry is another example. One of the glamour of Chinese ancient poems is that they are concise and generalizing and meaningful. They are never redundant. The appeal of language is just like music, it is the art of pause. Poem is just an outstanding example to demonstrate its characters. With its precise and concise words, it creates a meaningful and thoughtful artistic conception. Capture the imaginations of its readers. The words and phrases are just like a painting. Here words and phrases are the media, they convey the wordless spirit. It is so-called "to express the infinite with limited words" Liu Xizai once said in his book *On art and poem* "the wonder of poem lies in,its pause". So, literal meaning might be the opposite in the mind of the writer. Poem is a good example of mixture of reality and spirit. Wangwei in his poem once wrote"Straight lonely smoke rises in the desert, Grand long river reflects the round sun sets." The word "straight" and "round" is literally a description of shape, it actually imply an scene of unmeasured vastness. the gentle and breezeless the clear and bright scene reflects the desolate and loneliness. Chinese poem of four lines is intended to limit the characters in the poem, the strict limitation not only means tidy and precise, but also reflect the value of spirit. There are some differences compared with painting. There are countless instances such as the art of iconography, Ming–style furniture, calligraphy and drama. Then, let's go back to architecture.

Architecture is so real and mighty. It erects in any site to display its solid and dimension. In modern China, architecture on one hand makes contribution to the

调了"言"的规整与严谨的同时，反衬出"意"的可贵与难得，由此与绘画有很大区别。除此之外，我们还可以举出无数例子，如造像艺术、明式家具、书法以及戏剧等等。然而，回到建筑中来吧。

建筑这个东西是如此的实在又强势。它矗立在场所中，炫耀着自己的坚固和尺度。在现今的中国，它又成为经济发展中，城乡一体化进程的功臣与资源消耗、环境恶化的帮凶。作为这一代的有责任感的建筑师，我们应该如何应对呢？先看看我们的老祖宗是如何做的吧，古代的中国是讲"天人合一"的，由于技术的限制，建筑尺度除了宫殿与庙宇外，都不是太大，而空间关系成为建筑者发挥才能与想象的地方。最集中的例子就是园林建筑了。园林建筑的理论分析有很多了，包括我自己也写过类似的东西，这里我们就前文"有无相生"的角度来看看园林建筑吧。园林建筑是不能孤立地看待的，它是与中国诗、书、画等艺术形式相结合的，我们刚刚就诗、画的生无之境做过描述，证明"有无"的意境，特别是"无"的意义之重大。而建筑是如此之实，与语言、书、画有极大不同，先人们又是如何处理的呢？有趣的是，他们用的同样是很实的东西——"墙"，尤其是白墙。建筑是实的，但建筑里面是空的，外边也是空的，建筑与建筑之间还是空的，而所有的空只是因为一墙之隔。墙成为空间的媒介，而就此角度来看，又类似诗与画了。看来任何事物都是决定于角度的不同。先人们把墙当做了绝句的韵律来看待，墙与其规则、工法与材料习惯，而墙的实或"有"，成为空或"无"的有力的衬托。同时墙有其特有的特质，可以附着很多东西，如光与影，如诗与画，如雨与渍。同时也可以去掉很多东西，如窗与洞，如门与扇。这些附着的和去掉的都是为了墙以外的东西。光影带来墙的时间性与运动性，甚至生命性。什影的斑驳，诗意地带给白墙无限的生机。诗与画令白墙以更实的存在，证明了其相对意境的不存在，其自身的物质属性的虚无。雨水的冲刷留给白墙的是浓淡的渍，如水墨般的偶然性与历史的必然性形成了"有无"的对话。去卓的部分成为门与窗，而门与窗的位置是那么的特别，它既是虚的洞与周边实的"有无"对比，又是尺度与密度（数量）上的"留白"。从中感到先人们如绘画般寻求墙中的"气"，以及"气"与周边场的关联。而最重要的是洞以外的世界透射进来，这时相对墙的"虚与无"，变成了更为有意味的另一种"实"与"有"。外面的空与里面的空有了对话的渠道。这时的门与窗不再只是通行与采光的功能构件，而是借景与对景的道具，空间融通了，情绪涌动了，氛围来到了。这就是我们的前辈建筑家造就的神话般的传奇。运用了原本无比实在的手段——"墙"，成就了建筑的灵性。

墙同时还可以成为连续空间的媒介，比如"进"。中国建筑的精髓在于两方面，一就是园林与民居了，另一个就是宫殿与庙宇。但无论哪一种，都讲究"进"。一进院两进院的"进"。墙在这时表面上退化到了界限的范畴。实际上还是"进"与"进"之间的"有"。而"无"的就是院落了。院落的氛围各有不同，而墙决定了它的基本性格，有了墙才有了园林，才有了更多的附着物。太湖石、水池、亭台、竹林等等。而这时的墙成为了庭院与庭院之间的过渡，室内与室外的过渡。室外景观成为室内墙面开窗部分的理由，太湖石变成室内窗洞后的装饰画，窗成为了取景框，最终墙成了沟通的媒介，空间的延展，也就由原本的局限物变为拓展物。

中国园林里还有一种墙，是白墙的兄弟，叫"格栅"。顾名思义，是木格子的隔栅。它是墙的延展，更具通透性，格子本身就是抽象的构成装饰物。同时也是空间划分与限定的手段，只是它更虚、更轻、更灵动、尤其是

integration of city and countryside. On the other hand it plays a disgraceful role in resource consumption and environment deterioration. As a responsible architect of the contemporary society, what should I do? Firstly, let's review what our ancestors had done. In ancient times, Chinese paid much attention on the harmony between people and nature. As for the limitation of the technology, the dimensions of the architecture were not too large except palaces and temples, so the space was the place where the architect could display their talents and imaginations. The typical example is the landscape garden architecture. There are so many kinds of analysis theories of the garden architecture, including those composed by me. Followings are the analyses of the garden architecture from the viewpoint of reality and spirit. The garden architecture should not be analyzed isolatedly, for it is the combination of the Chinese art forms such as poetry, calligraphy and painting. As described above, let's start from the viewpoint of mixture of reality and spirit, to give an overview of Chinese architecture. The architecture is so real and differ from the language, calligraphy and painting. How did our ancestor to deal with it? Interestingly, they use the solid "wall", especially the white wall to play the role as a blank. The building was solid, but the inside and outside were empty, the distance between the buildings were also vacant. All the rooms and spaces were generated by the wall separation. Wall is the medium of the space. From this point of view, it has some similarities with poem and painting. So the different points of view made all the differences. The ancestors regarded the wall as the rhythm of a four-line poem. The wall has its shape, working method and material. The reality or the existence of the wall serves as a foil to the spirit or the inexistence. Meanwhile, the wall has its special characteristics, it can be attached to many other decorations, such as light and shadow, poem and painting, and rain and trace. On the other hand, it can be detached, such as window and hole, gate and door. All the attached and detached are for something besides the wall. The light and shadow bring the temporality, movability and even life to the wall. The mottled bamboo shadow endows the white wall vitality. And the poem and painting make the white wall more real and testify the inexistence of the relative conception and the material character of itself. The vain erodes prints slightly trace on the white wall, just like the "reality and spirit" "communication in a Chinese ink painting. The detached parts are door and window; however the location is so special that it is not only the reality and spirit comparation of the virtual hole and surrounded reality, but also the white space for the dimension and the density. We can catch the field in the painting of our ancestors, and the connection of the field and the surround. The most important is the word illuminated outside of the hole. By now, the scene will be another "reality "and "existence" scene relatively to the "visual and spirit" wall. Then there is

the channel for the communication between the outer and inner. The door and the window are not the functional structure for passing and lighting, but the tool for the view borrowing and view in opposite place. Then the space become light with the emotional impulse and the atmosphere will be tender and romantic. All the fabulous legends are what our predecessor architects had created. They used the most specific method——wall——to create the spirituality.

The wall may also become the medium connecting the space,the entrance, for instance. The essence of the Chinese architecture lies in two aspects. One lies in gardens and dwellings, while the other in palaces and temples. Whichever, they all stress the entrance. From this point of view, the wall is only for limits and boundary. And actually it is still the reality between the "entrances", while the "spirit" is the courtyard. The courtyards have different ambient, while the wall decides its basic character. With the existence of wall, the courtyard and some other decorations emerge, such as great lake rock, pool, pavilion, and bamboo grove, etc. Now the wall is the bridge for the courtyards, and for the indoor and outdoor. The interior wall window is opened for enjoying the outdoor landscape; the lake rock becomes the decorative picture of the indoor window, and the window plays the role of viewing frame. At last, the wall contributes to be the communication medium and the space. And the wall becomes the extension from the former limitation.

There is another kind of wall, named grating, which has some similarities with white wall. Grating is the cell barrier. And it is the extension of the wall and has more visual transparent. Grid is not only the abstract decoration but also the method for special division and limitation. Being more humble, lighter, and more living, the grid plays an invisible role, especially in interior wall. It not only has the advantages of the white wall, such as open the hole and gate, but also has the comparision of the reality and fantasies. And it also combines the light and shadow. When the light and shadow permeate through the grid, the shadow is dancing in the floor, and makes the interior romantic and mysterious. The grid overlaps the scenes in the form of the hollow construction, and makes people feel traveling in different times. Meanwhile, it combines with the sound and flavor. With the rustling rain in bamboo forest, the murmuring water flow sound,birds' song and fragrance of flowers, the grids constitute the subject of the space. The combination of the wall and the grid motors the sense of hearing and smell, besides the sight. The grid is made from wood, being more close to the invisible wall and the furniture. The touch feeling is so tender that it is mixed together with the grid. Finally, the wall and the grid accomplish the great achievement in perception,

内墙中，栅起到了灰空间的作用。它兼具了白墙的优点，比如同样可以开洞与门，同样有实与虚的对比，同样吸纳了光与影，只是光影透过它时，成就了室内的迷离，影在地上的移动，引来室内其他物品的嫉妒。栅以虚空的构成重叠了景象，变幻了时空。同时它还吸纳了声音与味道，竹雨沙沙，水流潺潺，鸟语花香，与影子的光怪陆离，组成了空间的主体。感官不只局限于视觉了，墙与栅调动了听觉、嗅觉。栅为木制，更为近人尺度的虚墙，更接近家具的功能。触感是温和的，所以触觉也被融合进来了。最终墙与栅成就了感观之大成。将原本死板的空间变活了，将情感融入了。"苔痕上阶绿，草色入帘青""谈笑有鸿儒，往来无白丁"。多么富有诗意的空间，而人最终成为主角，人令空间的灵性再一次质的飞跃了。这就是我们先辈建筑家的工作，充满了智慧与愉悦的工作，是我辈应当寻回的工作状态与态度。否则我们将成为强势建筑的俘虏，成为空间的奴隶，"有"与"无"本末倒置了。而本来是正相反的。它们早就存在在那里，等候我们多时了。

现在让我们回到现实，当代的建筑师拥了更多的技能与技术支持，我们可以说战无不胜，攻无不克地挑战着各个技术与空间难题。而创造的是都正确的吗？一次与肖老师聊天，他说中国正处在国家的青春期，年轻人都会犯错，犯错是正常的，不犯错是不正常的。所以不用为之气恼，长大了就自然会好。可话虽如此，难道我们就什么也不做，听之任之吗，季羡林老先生说，21世纪是中国人的世纪。他指的当然是中国的哲学与文化的发展空间。而在此之前，我们是单纯的拿来主义，如果我辈不在此时反刍一下我们自身的文化精髓，难道要若干年后再由老外来教我们怎样做园林建筑吗？他们会说，回家看看吧，我也是从那儿学的。我们当然要学习他们的优点，比如理论体系的健全，分析比较的方法，理性认真的态度，然而在他们正努力从我们东方的精神家园寻找突破口时，我们又有多少人意识到自己与生俱来的优势呢？再拿建筑来举例吧，我们建造的空间从来就不会是完美的，建筑本身就是遗憾的艺术。何况现在的更高更强是下一个更高更强的笑柄。中国的青春期充满了欲望，并且引发了世界的欲望，我们更需要了解自身的结构，调整内分泌，而不是纵欲过度吧。

有人说存在的就是合理的，那是指"有"的就是有用的吗？为了有而有，那只是占有欲，为了情感的共鸣，心性的融合才是爱情。建筑师应该和自己的建筑谈一次恋爱了。而感情就像水流，最终是要流向大海的。中国的建筑向何处去？我的答案就是"有无相生"，"有无之境"了。当我们面对项目时，需要首先在"无中生有"的角度看待它。场地就像张白纸，我们勾画的物体要如同"留白"一样建立存在"气"的场，令场所与周边对话，而往往应该从这个"场"来开始工作，而不是先从建筑的轮廓开始，建筑就是我们对话城市，历史与社会的媒介。我们表达的是我们对建筑与社会的责任与对建筑艺术的追求。无论是任何对象，都不一定要妥协它的符号背景。因为符号是一种低级的代码，就像画驴的与画猴的区别一样，而画家的主题不一定是动物，而是心性。看得到的建筑体都是我们的道具，材料是我们说话的单词，是音乐的音符，而词或音符之间的空间与停顿，是我们调动其特征产生为己所用的效果的工具。空间的营造要学习先辈们运用"墙"的智慧，"墙"的存在是为灵性的空间服务的，令空间的划分与限定变为以空间的互相联系与流动为前提的丰富性的创造工作。如同绘画追求的画面以外的联想与意境。相信当我们有了功能的合理布置能力的话，再以空间的表现力与我们赋予空间的精神性等这些并非一眼看得到的"无"为手段的话，我们很可能就不会为建筑的形式

which activates the inflexible space.It reminds me two poems. "The footsteps is crept over by the green moss, and the green grass jumps into the sight." "When talking, there are only great scholars without any innocent people." Such beautiful poems full of poetry space. Moreover, people is the protagonist, which makes the space spirituality a qualitative leap. This is the work of the ancestor architects, the work is fulfilled with intelligence and pleasure, the working condition and the working attitude we should follow. Otherwise we will be the capture of the mighty architecture and the space, which put the "reality" before the "spirit". The working condition and the working attitude created by our ancestors have always been there for ages.

Let's go back to the reality. The modern architects possess much more technical skills and support, in the process of overcoming all the technical and special problems. However, are all the created correct? Mr. Xiao said to me once in a conversation, China was in the national adolescence. Young people were easy to make mistakes, which was normal. No mistakes was strange. So it was no need to be annoyed, and it would become normal when it grew up. Even so, shall we do nothing about that? As Mr. Ji Xianlin, the famous linguist and social activist once said, the twenty-fist century was the century of Chinese. What he indicated was the develop space of the Chinese philosophy and culture. Before the 21st century, what we did was just bringing. If we would not ruminate our cultural essence, shall the foreigner teach us how to build the gardens years later? If any, the foreigners would say go back to your home, what I have learned is from your country. Absolutely, we should learn their advantages, such as the sound theory system, the analysis and comparison method, and the rational and earnest attitude. In the process of the foreigners are trying hard to seek methods from the oriental spirit, however we seldom realize our inherent advantages. Take the architecture as one example again. The built space will never be perfect, as the architecture itself is one woeful art. Moreover, the current high and strong building will be swallowed by the further higher and stronger building. The Chinese adolescence is full of desire, and initiates the world's desire. We should understand our structure and adjust, but not indulged in desire.

There is saying that what exists is reasonable. However, is existence really useful? For the purpose of being existence, it is the desire to possess. Love is for the purpose of communication of the emotion and the disposition of mutual understanding. Architects should communicate with their architectures. The emotion is just like water current, and flow into the sea at last.So where will the Chinese architecture flow into? My answer is that the existence and inexistence generate each other and reach the reality and spirit state. When we begin one project, we should observe it from the point of view of "the mixture of reality and spirit". The field is just like one piece

感而困惑了。她就在那儿，她就应该如此。"得之于手而应于心"吧。

　　另外，地域性是我们努力要铭记的，中国建筑师快变成同质化的无聊的粉刷工了，到处都在无情地摧毁其地域特色。而我不是说地域性就一定要用或一定不用地域的特色形象或符号。因为我们讨论过园林的精要，首先地域性来自于人们的感观的总体，不单纯是视觉，我们要首先善于观察，我们建造的地点的特色与人文景观的深刻之处，然后就是我们选择承载它们的媒介的任务了。我们将全部感观的作用变成功能，去完成一种美，一种富有地域特点的并不沉重的闲情。用看不见的，但逐渐浓郁的地域氛围来填充空间的全部。当然适当的符号是可以用的，但也只是以有用无，它所传达的我们建筑师对建筑经过思考与分析得出的一部分意义，并非是对无奈的枯槁的粉饰。就如同你不学外语，也懒得去国外，但非要把自己整形成一个老外。看来很像，但除了一副躯壳外，只有谎称自己失忆了。而我们如果继续错误下去，就相当于自己对着镜子给自己做手术一样，最后只能集体失忆了。

　　当我们执著于很多东西时（例如形式语言、业主喜好、符号等等），跳出这存在的"实有"，不要执著于表象，提高自身的感悟能力，用艺术家的眼光看待对象，以技术与功能为"有"，并"因其所有，用其所无"，"得之于手，而应于心"的创造无限的"无"吧。

邹迎晞

2011年4月5日

结于昆明

of white paper, the objects that we draw are just like the white space forming one kind of field atmosphere, and make the field communicate with the surrounding. So it should begin from the field, not the architecture outline. The buildings are the medium we communicate with the city, history and society.

What we express is the responsibility for the architecture and the society and the pursuit of the art of architecture. For any object, do not care much about the symbols. Symbol is just one kind of basic code; it is just like the difference in donkey and monkey painting. The subjects of artists are, maybe, not the animals, but disposition. The visual buildings are the tools, and the materials are the words and the musical note, while the pauses between the words or notes are the tools used to display the features in order to produce the required sound effect. We should learn from the wall of our ancestors to make the space wisdom. The existence of the "wall" serves for the spiritual space, and makes the division and limit of the space become the rich creation work with the premise of interconnection and fluxion of the space, which is just like the case of pursuing the off-screen imagination and conception in painting. It is believable that after we master the reasonable function arrangement ability, and the invisible methods such as the ability of represent and the spirit we endow with the space, we will not be confused by the architecture form probably. Master the knowledge, and then apply it with efficiency.

Besides, the regionalism is what we should keep in mind. The Chinese architectures almost become the boring plasterer, and destroy the regional characteristics ruthlessly everywhere. What I mean is that the regionalism is not the necessary for the regional image or symbol. As what we have discussed about essence of the garden, the regionalism come from the people's overall feeling, not from sight alone. First we should expert at observation on the profound significance of the unique feature and human landscape of the building place. Then, we should choose the medium to display the above significance. We transform all the perception into the function to accomplish one kind of beauty and one kind of quietness full of regional characteristic without lumbersome. We use the regional invisible but gradually intense atmosphere to fulfill our space. Nevertheless (Of course), we can use suitable symbols in some cases, which are not the decoration of paleness. They can convey a part of meaning through thinking and analysis by architect. The situation is just like you do not like to study foreign language, nor go abroad, but you got plastic surgery in order to get an appearance of a foreigner, then you have to lie about memory loss as you got nothing more than a similar appearance. If we commit our fault repeatedly like that, it is just like we operate on ourselves in the mirror. The final result would be we all lost our memories after we got the collective amnesia (memory loss).

When we are insistent in something such as formal language, the owner's favor and symbol, please let us step out of the exist "reality", do not be persistent on representation anymore. Improve the apperception ability, and enjoy the object in the view of artist, regard the reality as the technology and performance and create the boundless "spirit" by the means of "base on the reality, utilize the spirit";and "master the knowledge, and create the spirit in the limitless realm".

By Zou Yingxi
April 5, 2011
In Kunming

从左至右: 贝抚林[德]、曹荣平、福客[德]、邹迎晞、高杨
From left to right: Florian、Cao Rongping、Volker Kunst、Zou Yingxi、GaoYang

后记

　　我在书还没有构思好，甚至还没有回北京，就开始写后记了，甚至在写前言之前。这是因为博物馆还没有完工，这个早产儿就呼之欲出了。还有那么多工作没有完成呢，很多空间的细节，材料的处理，立面的收口，内装修的补救正在进行中，或正在浮现出来。还是一周前，我还在郑州与王迪馆长及肖老师定下了立面石材的材质与规格。室内设计正在进行中，这次肖老师同样发挥了巨大的作用，虽然他自己就是优秀的室内设计师，他还是明确的支持王迪馆长由我来继续完成。但他提出了高明的见解，并及时的纠正我的思维。本以为经过建筑设计阶段已经得到翻天覆地的变化并进步得一塌糊涂的我，又一次惭愧的汗颜了。因为我又一次犯了"执"的毛病。那句"旁观者清"真是至理名言啊。肖老师不亲自执笔来创作也是因为他自知也会"执"吧。

　　所以写完理念阐述这最累人的一段后，我想还是理一理没完成的工作，顺便检讨一下自己。因为我在写上一段文字时，又犯了"执"的毛病。好的理念应该不用写那么多，那么狠吧。自己是不是又太想把话说完呢？既然写了，我也就不想再改了，好不容易有自己做主的机会。不过如果真的再写一遍的话，我会把它分成两篇，一篇是理念阐述，言简意赅，另一篇就是设计随感，扬扬洒洒了。其实想改也没机会了，没时间了，何况早以发回北京送去翻译了。碰巧自己又必须留在老挝几天，有能量的话，再赶出一两篇文字吧。只是可惜了这迷人的异国情调，碰到我这个"身在曹营心在汉"的老外。老挝碰巧是个佛教的国度，让我在这古老的佛教王国里梳理龙门博物馆的设计思路，也是一种巧妙的安排吧。

2011年4月2日
于老挝万象市

Postscript

I started writing this postscript even before I conceived my book and returned to Beijing, particularly before writing the foreword. This is because the book is urgently needed before the museum is completed. However, there is still a lot of work to do, such as space details, treatment of materials, close-up of façade and remedies of interior decoration. They're underway or emerging. A week ago I determined the materials and specifications of façade stones with curator Wang and Mr. Xiao in Zhengzhou. The interior design is also in progress and Mr. Xiao still plays a vital role in it. Though he's the best interior designer, he clearly suggests to curator Wang that I continue completing the entire work. Moreover, he provided me with many valuable suggestions and corrected my thinking in a timely manner. I originally thought that I had changed my mind thoroughly and made unbelievable progress, thus I felt deeply shamed again this time. I made a mistake because I was clinging again. Just as a proverb says, "The onlookers see most clearly." Maybe the reason why Mr. Xiao didn't create it in person was because he feared he would also be clinging.

So after writing this most tiresome paragraph——concept interpretation, I guess I had better sort out what to do and reflect on myself because I was clinging again during my writing last paragraph. For good concepts, maybe they shouldn't have been written. Am I too eager to express? I didn't want to amend it now that I wrote because it's hard to take the opportunity to control your own fate. But if I'm required to write it again, I'll divide it into two articles, one is concise concept interpretation and the other is casual design impression. In fact, I have no time and no opportunity to amend it because it is already sent to Beijing for translations. I happen to stay in Laos for a few days and will write one or two articles if I'm energetic. It's a pity that I can't appreciate the charming exotic scenery. On the other hand, Laos also happens to be a Buddhist country. Perhaps it's a subtle coincidence that I should be able to groom my design ideas in such an ancient Buddhist kingdom.

Zou Yingxi
April 2, 2011
Vientiane, Laos

图书在版编目（CIP）数据

古阳建境 / 邹迎晞编著.—— 北京：中国书店,2011.4

ISBN 978-7-5149-0057-6

Ⅰ.①古… Ⅱ.①邹… Ⅲ.①博物馆—建筑设计—洛阳市 Ⅳ.①TU242.5

中国版本图书馆CIP数据核字(2011)第056331号

编　　著：邹迎晞

编　　辑：李云峰

设计制作：李云峰　姜丽楠

出　　版：中 国 书 店 (北京市西城区琉璃厂东街115号)

网　　址：www.zgsd.net

印　　刷：北京恒嘉彩色印刷有限公司

版　　次：2011年4月第一版　2011年4月第1次印刷

开　　本：880×1230　1/16

印　　张：11.75

印　　数：1——2000册

定　　价：198.00元